The Korean Economies

The Korean Economies

A Comparison of North and South

EUI-GAK HWANG

CLARENDON PRESS · OXFORD
1993

Oxford University Press, Walton Street, Oxford OX2 6DP
Oxford New York Toronto
Delhi Bombay Calcutta Madras Karachi
Kuala Lumpur Singapore Hong Kong Tokyo
Nairobi Dar es Salaam Cape Town
Melbourne Auckland Madrid
and associated companies in
Berlin Ibadan

Oxford is a trade mark of Oxford University Press

Published in the United States
by Oxford University Press Inc., New York

British Library Cataloguing in Publication Data
Data available

Library of Congress Cataloging in Publication Data
Hwang, Eui-Gak.
The Korean economies : a comparison of North and South / Eui-Gak
Hwang.
p. cm.
Includes bibliographical references.
1. Korea (South)—Economic policy. 2. Korea (South)—Economic
conditions. 3. Korea (North)—Economic policy. 4. Korea (North)—
Economic conditions. 5. Korea (South)—Relations—Korea (North)
6. Korea (North)—Relations—Korea (South) I. Title.
HC467.H79 1993 338.9519—dc20 93-23769
ISBN 0-19-828801-8

1 3 5 7 9 10 8 6 4 2

Typeset by Graphicraft Typesetters Ltd., Hong Kong

Printed in Great Britain
on acid-free paper by
Biddles Ltd,
Guildford & King's Lynn

*This book is dedicated to the Lord Jesus Christ
who cares for His people always,
and who is the foremost supporter,
moral economist,
and political adviser
in my life*

Acknowledgements

In 1990 I was invited by the Brookings Institution in Washington, DC, to conduct an independent study on the economics of North and South Korea. The book was written mainly while I was with the Brookings think tank.

I received valuable and constructive ideas from many people who participated in the seminars, held every Thursday. I am particularly indebted to Dr John Steinbruner and Bruce MacLaury for all their generous help and encouragement despite their very busy schedules. I am also very grateful to those experts who extended to me many valuable and constructive comments—Professors George Tolley, D. Gale Johnson, and other participants in the Development Economic Workshop at the University of Chicago. John Steinbruner, Harry Harding, John Merrill, W. Robert Warne, Yeon H. Choi, Se-Kon Choy, Yul-Hong Min, Un-Bong Chung, and other participants in the Foreign Policy Studies Program's seminars at the Brookings Institution gave me valuable ideas and suggestions for improving many critical parts of the book. I also greatly benefited from thoughtful and useful comments of Young-Ja Kang, Dr Il Sakong, and Professor Lawrence B. Krause and another, anonymous, reader.

Several individuals provided me with excellent administrative and research help. In particular, Charlotte Brady, Susan Stewart, and Susanne Lane gave me every friendly assistance while I worked on the manuscript, and Judy G. Buckelew toiled in typing and retyping the whole manuscript with many stylistic improvements.

Many readers and editors at Oxford University Press helped me refine the contents: Andrew Schuller, Enid Barker, and Sue Hughes merit particular mention. But of course, any remaining errors and shortcomings are solely mine.

E.-G. H.
Seoul, February 1993

Contents

List of Figures

List of Tables

Introduction

The objective of this book is to analyse and compare the parallel developments of two rival economies, North Korea and South Korea. Against the common historical, social, cultural, and linguistic homogeneity, the North chose essentially an autarkic socialism, while the South followed the road of market-oriented capitalism. The two states have not interacted substantially for almost forty years because of mutual distrust and hostile strategies arising from the three-year civil war in the early 1950s. Undeniably, the Korean War brought untold sorrow to the nation, deepened mutual enmity, and was detrimental to national cooperation; this is the reason why Korea is now the only divided country left in the world.

After the 36-year colonial rule of Japan, Korea (the 'Land of the Morning Calm') was liberated with the defeat of Japan at the end of the Second World War in 1945. While the Korean people immersed themselves in joyfulness at the nation's independence from Japan, the two superpowers—the United States and the Soviet Union—each occupied half of the country under the pretext of facilitating an acceptance of the Japanese surrender. Under their auspices, eventually two rival systems were installed—a capitalist South and a pro-Soviet communist North.

The division of the country by the external forces extends through all phases of Korean life—political, economic and social. The artificial division at the 38th parallel destroyed national unity, led to tragedy of fratricidal war and produced many other human and national agonies. The division broke the mutual complementarity between the northern part and southern part of the economy. For example, before the division northern Korea had almost 70 to 80 % of the heavy industry including most sources of hydro-electrical power supply, while southern Korea dominated most of the light industry as well as rice production. The consumer-goods industry of the south depended upon northern supply of both electrical power and raw and semi-finished materials. The south depended on the chemical fertilizer produced in the north, while the north needed the rice from the south.[1]

In this way, northern and southern Korea complemented one another, and each was highly dependent upon the other for the satisfactory operation of the country's overall economy. The division made the two halves eschew this complementarity and instead enter into hostile competition, rooted in fundamentally opposing economic systems. The South follows a capitalist market economic system, while the north maintains a socialist command economy, in which all means of production are owned by the state and central planning organs plan and direct the overall economy.

Thus, the North and South Korean governments have launched mutually exclusive economic development plans ever since the end of the Korean War (1950–3), and now claim to have accomplished much progress in their respective economies and systems. South Korea maintains that it has achieved a remarkable economic growth and development in the environment of a mixture of free-market system and government indicative planning over the past three decades. Meanwhile, North Korea claims to have successfully built the model communist nation. North Koreans try to convince the world that the majority of their compatriots in the South suffer from hunger, homelessness, and exploitation under the non-egalitarian capitalist system. South Koreans maintain that their compatriots in the North are being fed only two meals daily because of reduced food production and rationing. All these self-serving and often false statistics and rumours put out by each side serve to obscure the world's perception of both Koreas. Verification of such uncertain information requires some objective and consistent comparative analysis of the North and South Korean economies. However, this task is extremely difficult because of the limitation of comparable or equivalent data-sets of the two rival systems as well as the differences in economic systems.

South Korea's capitalist policy has resulted in abundant production of both consumer and exportable goods over the past three decades. In recent years, however, the South has begun to experience unforeseen social conflicts between the haves and the have-nots, and there is concern over the deteriorating morality among the people and public servants which are by-products of an emerging selfishness and materialism.

North Korea, on the other hand, has adopted a *chuche* ('self-reliant') ideology of autonomy, effectively isolating itself from much of the world. The communists appear to have achieved a somewhat egalitarian society in which the masses are motivated primarily by the principle of serving the common good rather than a concern for self-aggrandizement. However, many economic difficulties appear to have emerged more recently in the North, probably as a result of the lack of profit incentives and the increasing complexity of the communist economy: there has been a diminishing marginal return to the forced social drives in the country's efforts to increase production. Since the early 1970s, in fact, North Korea seems to have experienced both a lowering of morale and a decline in productivity. The stepped-up *chuche* indoctrination, and the increase in slogans exhorting production increases, can be understood to reveal such drags. However, there are conflicting reports about whether or not the North's economy is still robust. The aim of this study is to provide some objective basis for evaluating the North Korean economy in comparison with that of South Korea.

One contribution of this study is that a full set of time-series data on both North and South Korea has been collected, tested, and reconstructed, where appropriate, so that one-to-one mapping between the North and the South has been possible. The dearth of North Korea's data has been attributed to many different and often conflicting estimates of its economy. For this book I have attempted to collect consistent data-sets of the North so as to enable a comparison with the South in the framework of positive economics. Another important goal is to compare the two economies with a view to possible economic co-operation in the future, perhaps leading to economic integration some day.

Before delving into the mainstream of the comparison, Chapter 1 describes in brief a history of Korea from its origins to the Korean War. After that time the two halves of Korea have followed different paths of economic development—the subject of this study.

A comparison of the two economies begins in Chapter 2 by giving much attention to the historical evolution of economic development plans and policies of the two competing systems. It must be stressed that this book is for all readers who are interested in either or both of the two economies. The aim of Chapter 3 is to compare the two economies in terms of both a macroeconomic index and living standards. Indeed, while addressing the difference in ideological and economic systems of the two states, the chapter's most enduring contributions are, in my view, methodological: it seeks to set up a consistent GNP data-set out of the socialist GVSP (gross value of social products) and income estimates. Even if we ignore at the outset the differences between North and South in initial endowments of natural resources and other geographic factors, such as arable land and climate, there are several hazards involved in the comparability of their economic growth. Gerhard Breidenstein and W. Rosenberg pointed out the following problems:

First, socialist countries do not use those GNP aggregations which are common in all capitalist and most other non-socialist economies. In capitalist economies, GNP at market prices is the total value of all final products and services produced during a given period of time. Socialist economists, however, add the output values of all separately enumerated production units. The Gross Value of Social Production (GVSP) includes also the values of intermediate products, namely material costs for consecutive production units, thus making their value counted more than once.[2]

The problems entailed in obtaining comparative aggregate indexes, namely the GNP and income of North Korea, are discussed in great detail. To provide the ultimate empirical foundation for a rational selection among alternative estimates, both the principal-component and correlation analyses are employed. Living standards are documented comparatively for both countries, making the best use of available data and supporting materials.

Chapter 4 serves to introduce the historical development and existing structure of public finance in both economies, and to examine state (both central and local) expenditure distribution between military and other uses. For the South, some empirical analysis on the effects of government spending versus monetary policy is conducted. A similar study for the North is not, however, undertaken. One reason for this omission is that state spending accounts for a large part—about two-thirds of GNP—in North Korea.[3] This in turn suggests that there is a high correlation between state expenditure and the People's Economy, if we could set aside the efficiency question of government spending. Another reason is due to the lack of availability of relevant statistics such as money supply, prices, and interest rates. Many important and interesting questions concerning the effects of such public expenditure on economic well-being are left unanswered for the North, but the comparison of budget allocation between, e.g., defence and other categories is interesting.

Chapter 5 takes up the two Koreas' external transactions and their respective foreign policies. One point to note is that the volume of trade among communist nations usually remains at the minimum level required just to supplement any gap between domestic supply and demand. Such trade as there is reflects natural resource endowments, and so it has been with North Korea's foreign transactions. In other words, comparative advantage based on technology and labour costs does not play an important role except for military technology in the communist external trade decision. Hence trade is very small. In contrast, South Korea has aggressively pursued an export-oriented strategy, keeping economics largely separate from politics. It is not surprising that the South's policy of foreign dependency has recently proved increasingly more effective than the North's policy of self-sufficiency, as is globally evidenced in the economic growth race between the capitalist free-trade regime and the communist bloc's autarkic regime.

Chapter 6 records the beginning of contacts between the North and the South and discusses future prospects, which are, of course, very volatile, depending upon the changing circumstances around the Korean peninsula. On 17 September 1991 the two Koreas became full members of the United Nations. The entry into the UN will serve as a starting-point for the lessening of confrontation as well as for the beginnings of mutual economic and political interaction between North and South. The discussion in this chapter is rather descriptive, but the prospects of economic and social transactions are considered in view of existing and potential comparative advantages of the two economies, taking into account the reunification process and interrelations of the two Germanys. Professor Lawrence B. Krause, in his very thoughtful comments on my manuscript, suggested that I examine several models for possible reforms anticipated for North Korea: the China model (*perestroika* without *glasnost*), the USSR model (*glasnost* without *perestroika*), the Polish and Czech models (systemic collapse of the system), and the East German model (systemic collapse with external capitalist take-over). Such artificial categorization of different models is no longer persuasive, as the former Soviet Union (USSR) model is so mixed up with others, just as other types of models listed above have evolved over time. Nevertheless, I think the Krause suggestion would be very interesting if someone were able to predict how the North Korean regime would change, but I have left the important 'political' (rather than 'economic') questions, as well as the discussion of the costs, benefits, and problems accompanying reunification, for a separate study. Reunification is, of course, the most important policy issue at this time, but Chapter 6 aims only to address possible areas of economic transaction, such as trade and joint ventures between the two Koreas, which may work, in my view, as a solid foundation on the road to reunification.

Lastly, Chapter 7 deals with North–South Korean relations in the context of north-east Asian economic co-operation towards the year 2000. This chapter covers the changes in the relations of North Korea with Japan and the United States, and in the relations of South Korea with China and Russia. The security and the cost of Korean reunification are also discussed.

Nobody can predict when Korean reunification will become a possibility. With the Cold War coming to an end, Koreans see no reason why their country should be left out of the surging tide of *détente*. Nevertheless, the creation of harmony between North and South may require a great deal more time and patience from both parties.

NOTES TO INTRODUCTION

1. Bong-Youn Choy, *A History of the Korean Reunification Movement: Its Issues and Prospects* (Peoria, Ill.: Bradley University, 1984), 14.
2. Gerhard Breidenstein and W. Rosenberg, 'Economic Comparison of North and South Korea', *Journal of Contemporary Asia*, 5:2 (1975), 169.
3. In fact, South Korean officials employ the simplest method in the calculation of the North's GNP: they divide North Korea's official budget outlays by a factor of 0.6, or 2/3 its annual budgetary spending.

1

A Brief History of Korea

Across the ferry
By a path through the corn
Like the moon through the clouds
The wayfarer goes
The road stretches south
Three hundred li
Every wine-mellowed village
A fire in the evening light
As the wayfarer goes
Like the moon through the clouds.

'Wayfarer' by Park Mok-Gul (1916–78)

Throughout its 5000-year history, Korea has been trampled on by armies from China, Japan, Mongolia, Manchuria, Russia, and America. Yet it has often emerged like the moon from behind the clouds, and wayfarers the world over could again discover its unique cultural, social, and economic legacy—until its division in the wake of the Second World War.

The Korean peninsula protrudes southward from the northeastern corner of the Asian continent and is surrounded on three sides by sea. As is true of all countries, the geography of Korea has shaped its history, including the cultural, artistic, and philological legacy shared by its people. The Korean personality, which is sometimes out-going, ebullient, full of song and laughter, sometimes sombre and emotional, is the product of both the country's geographical location and its history.

1.1 THE ORIGINS OF THE NATION AND THREE KINGDOMS

Although its history begins considerably earlier than the seventh century, it was in AD 668 that Korea, as a unified country, came to occupy most of the territory it does today. Archaeology provides evidence of prehistoric man on the Korean peninsula, and historical linguistics reveals that the earliest Koreans were closely related to Mongolian tribes. However, as a historical entity with a cohesive culture and society, Korea began with the unification of three kingdoms (Paekche, Koguryo, and Shilla) by the Shilla kingdom in AD 668.

The period before the unified Shilla was the Three Kingdoms period. Koguryo was the first state in existence, in roughly the first century BC, and its territory extended far into what is now northeast China or Manchuria. The Paekche kingdom was only slightly younger and it occupied the south-western part of the Korean peninsula, which is known to have exported a lot of cultural traditions to the Japanese islands.

The Shilla was a latecomer, originating from the then Han federations. (Han, referring to Korea, is written with a different Chinese character from that used for the like-sounding 'Han' dynasties of China.) It occupied the south-eastern section of the peninsula and was to emerge as the strongest kingdom.

By the sixth and seventh centuries, the three kingdoms were locked into a triangular stalemate. Faced with the necessity of maintaining mutual independence, they formed alliances and shifted allegiances to prevent any one of the three from becoming too powerful.

Meanwhile, China, after its first unification (200 BC) by the short-lived state of Chin and then the long-lived state of Han, experienced a second unification by Sui (AD 640). The Sui dynasty sent major expeditions to conquer the Koguryo kingdom in Korea but failed, and this eventually was one of the reasons why the Sui state fell. The subsequent state, T'ang, was stable and long-lived. The T'ang dynasty dispatched three unsuccessful expeditions against Koguryo before a Shilla emissary sailed to China to forge an alliance with the idea of attacking first Paekche and then Koguryo. Shilla attacked by land from the west and caused Paekche to surrender to the united forces. After this victory T'ang withdrew, and 'occupied Paekche' came under the control of Shilla. Next, Shilla turned on its northern neighbour Koguryo while T'ang attacked in the rear. Koguryo, weakened by numerous previous battles with China, succumbed, and Shilla took over the control of most of the Korean peninsula south of the Taedong River. The northern third of defeated Koguryo territory came under Chinese control.

Thus, Shilla unified the three kingdoms of Korea in AD 668, but its reliance on China's T'ang dynasty had its price. Eventually Shilla had to resist by force the imposition of Chinese rule over the peninsula. This it did, but its strength again did not extend beyond the Taedong River. Much of the former Koguryo territory was given up to the Chinese and other tribal states. It remained for later dynasties to push the border northward to the Yellow and Tumen Rivers.

In this way, events of the seventh century largely delimited the territory of what was to become Korea. The Shilla unification marks the beginning of a unified Korean culture and language. Many new political, legal, and educational institutions were introduced during the unified Shilla period (AD 668–935). Domestic and foreign trade with both China and Japan prospered. Scholarship in Confucian learning, mathematics, astronomy, and medicine flourished. Buddhism, which first came to the peninsula in AD 372, was later accompanied by cultural refinements such as literacy in Chinese characters.

Shilla maintained its kingdom with a strong hand. Military commanders were stationed throughout the country, and a system of hostage-holding learned from the Chinese was established, whereby prominent tribal chiefs throughout the conquered territories had to send their eldest sons to Kyongju (the capital city of the Shilla kingdom) as a human guarantee that the tribes would not instigate rebellions against the central government.

Shilla, however, started to decline in the latter part of the eighth century when rebellions began to shake its foundations. By the latter half of the ninth century regional uprisings were no longer under the capital's control. Tax revenues fell off, leading to a further weakening of the state. The rebels in the former Paekche area declared the state of Later Paekche kingdom, and Later Koguryo was also declared. The chaotic situation eventually led to the emergence of the Koryo dynasty in AD 915 under a former general named Wang Kon, but Shilla did not cease to exist until AD 935.[1]

1.2 THE KORYO DYNASTY (915–1391)

The Koryo kingdom derived its name from the Later Koguryo kingdom from which the 'later' and the 'gu' syllables were dropped.

Although it had changed twice in its own language, this was the name by which Korea first became known to the western world and by which it is still referred to in the West today.

The founder of Koryo and his heirs consolidated control over the peninsula and strengthened political and economic foundations by closely following the bureaucratic and land grant systems of China. The Koryo dynasty witnessed nearly a century of thriving commercial, intellectual, and artistic activities parallel to those of China's Song dynasty (AD 960–1279). Stimulated by the rise of printing and publishing in the Song dynasty of China, the Koryo dynasty made a great advance when it invented movable-type printing in 1234, although Gutenberg is credited with its invention in Europe in the fifteenth century.

Particularly notable in the arts was the fine ceramic ware called Koryo celadon, a ware with a bluish-green glaze and fine high-fired porcelain body. Also remarkable was the world's most complete set of Buddhist scriptures, carved in wooden blocks for subsequent generations to use and re-use, which is now in store at the Haein-sa temple near Taegu city.

Koryo, with its capital in Kaesong (located in the northern part of Panmunjom on the 38th Parallel), became an active trade centre having transactions with China and other countries as far away as the Arab world.

While early Koryo was a glorious period, with flourishing cultural innovations, later Koryo came to experience growing external threats of war and conquest. Genghis Khan of the Mongol Empire dominated all mainland Asia. The Mongols launched a massive invasion in 1231. The Koryo armies put up fierce resistance but were no match for the highly organized mounted troops from the north. The Mongols subjugated the Koryo kingdom and reinforced their control through the intermarriage of the royal families.

Only in the early fourteenth century, when the Yuan dynasty of the Mongol empire began to fall and the new Ming dynasty of China was founded, did Koryo regain its independence. When the Mongols retreated back to Mongolia and the Ming dynasty established a garrison in the north-eastern Korean peninsula, the Koryo court was split between pro-Ming and pro-Mongol factions. In 1388 a general named Yi Song-gye, who had been sent to attack the Ming forces in Manchuria by his Koryo king, revolted at the

Yellow River and turned his army against his king in Kaesong. Yi took the throne in 1392 and founded a new dynasty, which was named Chosen (Land of the Morning Calm).

1.3   THE YI DYNASTY (1392–1910)

From the Three Kingdoms down to the Koryo dynasty, Buddhism was the dominant, and often the official, religion. Thus, Korean scholarship and arts owed much to Buddhism and to the Chinese culture which transmitted it to the peninsula. But the new dynasty founded by General Yi Song-gye's coup imposed many drastic reforms on the society. Among these was the most important reform of land ownership.

Land had been largely in the hands of Buddhist temples and monks. The fact that Buddhist monks had wielded overpowering influence in politics and economics, and thus had become corrupted by power and money during the latter part of the Koryo dynasty, led to strong reactionary measures by Yi Song-gye and his followers. Although many of the outstanding temples were permitted to remain intact and a few of the monarchs in the Yi dynasty were devout Buddhists, Buddhism's influence over the social fabric of society during this time decreased greatly, as they lost vast amounts of land.

The Chosen (Yi) dynasty introduced Confucian social norms and moral values which were to govern the kingdom for the next five centuries. But differences in the interpretation of Confucian ritual and etiquette produced factional groups among civil servants and scholars who began vying for power. The most productive period of the Yi dynasty came under its fourth king, Sejong (1418–50). King Sejong was noted for his mastery of Confucian learning. His wise politics were marked by progressive ideas in administration, phonetics, national script, economics, science, music, medical science, and humanistic studies. One of his most celebrated achievements was the creation of the Korean alphabet, *han-gul*, which became the official way of expressing the language of the Korean people from that time onwards.

After Sejong, however, the dynasty saw many ups and downs with endless conflicts, rebellions, quarrels, and power games within ruling classes and between people from different groups of discontented scholars, petty officials, ex-officials, and commoners.

Christianity, accompanied by Western culture and ideas, reached Korea through China in the seventeenth century. By 1785, however, the government had become incensed at the rejection of ancestor worship by Roman Catholics and therefore banned all forms of 'western learning', branding them as ill-bred, vulgar philosophy. Western ships from Britain, Germany, France, and the Netherlands began to reach Korean shores after 1801, seeking trade and other contracts, but the government rejected all contacts from abroad. The Chosen (Yi) dynasty pursued a completely isolationist policy against western 'barbarians' until late in the nineteenth century, earning the nickname of 'the Hermit Kingdom'.

In the meantime, the Yi dynasty was subjected to the battle for hegemony and expansion in East Asia between Japan and China, and later between Japan and Russia in alliance with France and Germany, in the late eighteenth and early nineteenth centuries.

The strategic rivalry between Russia and Japan exploded in 1904–5 and Japan won the war that ensued. Under the peace treaty of Portsmouth, signed in September 1905 through the mediation of President Theodore Roosevelt, Japan's 'paramount political, military and economic interest' in Korea was secured with Russian acknowledgement.[2]

Even before the conclusion of the Treaty of Portsmouth, the United States and Great Britain had already approved Japan's domination of Korea as a means of preventing Russian expansion in eastern Asia. It was the US President's view that realism demanded the sacrifice of Korean independence, and that a Korea controlled by Japan was preferable to a Korea controlled by Russia.[3] Thus, on 29 July 1905, the US secretary of war, William H. Taft, concluded a secret agreement with the Japanese Prime Minister, Katsura Taro, through which the United States approved Japan's complete domination over Korea in exchange for the pledge that 'Japan does not harbor any aggressive designs whatever against the Philippines'.[4] The British government also gave tacit consent to Japan's intention of establishing a protectorate over Korea: Lord Lansdowne's dispatch to the British ambassadors in Russia and France contained the statement that Korea, 'owing to its proximity to the Japanese Empire, its inability to stand alone, and danger arising from its weakness, must fall under the control and tutelage of Japan'.[5]

With this international approval, the Japanese government forced the Korean government to sign a protectorate treaty, the Portsmouth Treaty, with Japan two months later. The Chosen king and a majority of his cabinet members at first refused to sign the treaty, but when armed Japanese soldiers were placed around the palace, they had to sign at gunpoint. Thus, Chosen became a protectorate of Japan, and Ito Kirobumi, one of the Meiji ruling oligarchs, assumed the post of resident-general in February 1906, as *de facto* ruler of Korea. Thereafter a group of Korean intellectuals tried hard to promote educational and reform movements to make their people aware of the Japanese intention to invade, but by then Japanese dominance in Chosen was a reality. Japan annexed Chosen as a colony on 22 August 1910 and fully exploited it to feed the Japanese war machine until 36 years later, on 15 August 1945, when Korea recovered its independence after the Japanese defeat in the Second World War.

1.4 JAPANESE COLONIAL EXPLOITATION (1910–1945)

> The Land is no longer our own
> Does spring come just the same
> To the stolen fields?
>
> *'Does Spring Come to Stolen Fields?' by*
> *Yi Sang-Hwa (1900–43)*

The Japanese economic exploitation of Korea began with the conclusion of the Portsmouth Treaty. Between 1905 and 1908, Japanese control of Korea's currency was secured with the rapidly growing quantities of Japanese Daiichi banknotes. With generous support and loans from their home government and banks, Japanese merchants could easily expand their activities in the Korean market. The number of Japanese residents in Korea increased from 126 000 in 1908 to 210 000 by 1911. And the Office of the Resident-General enacted a series of laws concerning land ownership to the extraordinary advantage of Japanese in Korea. In 1908, the Oriental Development Company was established to help Japanese farmers resettle in Korea and seize Korean land, including royal property of unreclaimed land and military farms. Within a year, the company had seized 30 000 hectares (75 000 acres) of military farms and unreclaimed land. Royal property was taken away from the royal household in order to minimize the power of financial management.[6] This was aimed at preventing Chosen's last emperor, Kojong, from gaining any power base for rebellion.

Development during the colonial period can be best understood by studying the requirements of Japan and its strategic colonial policy changes over time.

1.4.1   Agriculture

From the beginning of colonial policy, Japan recognized the potential for increased rice production in Korea and placed emphasis on agricultural development through an expansion of cultivated area, irrigation and drainage facilities, and improved material inputs. The aim was to increase output and hence exports to Japan.

One feature of Korean agriculture of that time was the strong concentration on the production of crops, especially food grains and, within food grains, rice, as shown in Table 1.1. Crops occupied 91% of net agricultural product in 1910–12, and 81% as late as 1934–6. Rice output accounted for more than half the value of production of the major crops throughout the colonial period.

Rice exports to Japan accounted for 16% of production in 1910–15, rising to 44% in 1930–6. More than half of Japanese rice imports, from 1925–6 to 1937–8 and, with the exception of 1939 (very poor harvest), probably into the war years as well, came from Korea.[7] The increase in Korean rice exports to Japan was made possible through increased production on the one hand (as shown in cols. (4) and (6) of the table) and exploitation of Korean farmers on the other. More and more Korean farmers were downgraded by the colonial policy to either tenants or semi-tenants; in 1931 they numbered nearly 12 million, comprising 2 325 707 households paying high farm rents and living in a state of near starvation. The farm rents, a principal means of exploitation, were as high as 50–80% of a farm's annual income. In addition, Korean farmers were chained by usurious loans from Japanese capitalists with very high interest rates.

According to statistics compiled in 1930, about 75% of Korean farming families fell into serious debt. More than 70% of these debts was payable to Japanese financial institutions, at interest rates of 15–35% a year. The only way to avoid starvation was to abandon the farm; indeed, many went to Manchuria or Japan,

TABLE 1.1 Measures of Korean agricultural production: annual averages, 1910–12 to 1939–41

	Market value, current prices (¥ million)			Production indices (1929–31 = 100)[b]		Rice export indices (1929–31 = 100)
	Crops	Other	Total	Total output	Rice	
	(1)	(2)	(3)	(4)	(5)	(6)
1910–12	288	30	318	65.7	67.4	4.0[c]
1914–16	356	54	410	83.7	82.1	21.6
1919–21	1034	102	1136	84.5	85.9	37.7
1924–26	987	147	1134	89.3	88.8	76.3
1929–31	618	141	759	100.0	100.0	100.0
1934–36	901	179	1080	104.9	110.8	118.6
1939–41[a]	1537	339	1876	125.3	142.8	—

[a] Production indices exclude the year 1939, in which the rice crop was exceptionally poor.
[b] Includes both agriculture and forestry output.
[c] Covers only 1911 and 1912.

Sources: (1)–(3), Suh Sang-Chul, 'Growth and Structural Changes in the Korean Economy since 1910', Ph.D. dissertation (Harvard University, 1966), table A-1; (4)–(5), ibid., table 111–13; (6), B. F. Johnston, *Food and Agriculture in Japan, 1880–1950* (Ann Arbor, Mich., 1953), 264.

only to find it no easier to settle there. According to the statistics of the governor-general for 1925, of all the farm deserters, 2.88% went to Manchuria and Siberia, 16.85% went to Japan, and 46.39% were scattered in the cities of Korea where they held marginal jobs.

It is evident that this combination of land taxes, high rents, and the extension of Japanese ownership, in addition to curtailing domestic consumption, was the means employed to provide rice for export to Japan.

1.4.2 Manufacturing

The Japanese policy for the manufacturing sector was initially oriented to prevent the development of Korean enterprise. The administration's tool to restrict investment in non-agricultural enterprise was the so-called Corporation Law (Chosen Company Regulations).

At the beginning of the colonial era, the share of manufacturing in total output was less than 7%, as shown in Table 1.2. However, when the Corporation Law was abolished in 1920 the manufacturing sector grew rapidly, with increases in the number of companies and the amount of paid-in capital.

The Japanese began to develop industries in Korea at around this time because of the country's advantages of cheap labour, its strategic location for north-bound expansion, and its abundant resources for hydroelectric power generation. Heavy industries (chemicals, fertilizers, steel, and war goods) were located mainly in the north, while light industries (textiles, printing, machine tools, and manufactured food) were established in the south. Table 1.3 shows the distribution of industrial structures in 1940 in the areas south and north of the 38th Parallel, on which line Korea was divided after its liberation.

The growth of manufacturing was accompanied by increasing concentration of output in large, Japanese-controlled firms during the colonial period.[8]

The industrial policy of the Japanese colonial government was oriented towards the transformation of Korea during 1931–42 into a logistical base for continental invasion. During this period, emphasis gradually shifted from foodstuffs production to such heavy industries as machines, chemicals, and metals. In 1939–41 the manufacturing sector occupied 29% of all industrial sectors, of which heavy industries such as machines, chemicals, and metals constituted about 46%. Production of agricultural commodities steadily declined in value, from 84.6% of the gross national product in 1910–12 to 49.6% in 1939–41.

It is interesting to note that agricultural output increased in spite of its relative share in industry dramatically falling (see Table 1.2 (*a*) and (*b*)). That was because the official enforcement of industrial development was accompanied by the colonial agricultural policy aimed at rice production. As the tide of war turned against the Japanese, they squeezed more and more agricultural products out of Korean farmers by means of 'kong ch'ul' or 'quota delivery'. Farmers were compelled to grow rice and use expensive fertilizers to fulfil their assigned quotas. Grajdanzev shows that average gross output per capita was about three times as high in

TABLE 1.2 Output growth, population increase, and industrial structure in Korea, 1910–12 to 1939–41

(a) Output and population indices (1929–31 = 100) (net commodity product in 1929–31 market prices)

	Agriculture	Forestry, fishing, and mining	Manufacturing	Total	Population[c]
1910–12	67.3	33.7	17.4	54.2[b]	66.0
1914–16	86.5	45.9	31.5	69.6	—
1919–21	88.6	46.2	59.7	76.9	84.5
1924–26	91.2	69.0	97.5	89.6	—
1929–31	100.0	100.0	100.0	100.0	100.0
1934–36	98.7	161.5	194.2	127.1	—
1939–41	117.3[a]	227.6	255.5	165.5 (155.5)	115.2

(b) Industrial structure (% share of net commodity product)[d]

	Agriculture	Forestry	Fishing	Mining	Manufacturing	Total
1910–12	84.6	5.3	1.9	1.5	6.7	100.0
1919–21	78.6	2.7	3.0	1.4	14.3	100.0
1929–31	63.1	6.6	5.8	2.2	22.3	100.0
1939–41	49.6	7.2	6.3	7.9	29.0	100.0

[a] Excludes 1939, an exceptionally poor rice year. The index for total output in parentheses includes 1939.
[b] Only 1911 and 1912.
[c] Based on 1 October census counts in 1920, 1930, and 1940.
[d] Based on current values.

Source: Suh Sang-Chul, 'Growth and Structural Changes in the Korean Economy Since 1910', Ph.D. dissertation (Harvard University, 1966), tables B-5 and II-4, stat. app., table A-1.

TABLE 1.3 Regional distribution of industrial structure in Korea, 1940

	South of 38th Parallel		North of 38th Parallel	
	Value of output (¥ million)	Proportion (%)	Value of output (¥ million)	Proportion (%)
Heavy industry	138	20	549	80
Chemical	91	18	411	82
Metal	14	10	123	90
Machine	33	69	15	31
Light industry	562	70	241	30
Textile	171	85	30	15
Processed food products	214	65	115	35
Others	177	65	96	35

Sources: Bank of Chosen, *Annual Economic Review of Korea* (1948); J. Shoemaker, *Notes of Korea's Postwar Economic Position* (New York, 1947); South Korea's Board of Unification, *Economic Comparison by Sectors between South and North Korea and Long-term Projections* (1972), 54.

Japan as in Korea in 1939, however.[9] This was due mostly to the difference in the occupational structures of the two economies; for example, in Japan a larger proportion of workers was concentrated in manufacturing and other relatively high-productivity sectors, while Koreans were forced to do non-skilled farm and heavy manual labour (at wages less than half those received by their Japanese counterparts). The expansion of Japanese colonial capital in the 1920s resulted in increased poverty and depression for the Koreans through increased exploitation, and their plight became a focus for the Korean national resistance struggle. It also stimulated the emergence of the socialist and communist movement that was in vogue at that time throughout the world.

1.4.3 Trade

Data on Korea's trade (exports plus imports) with Japan show an increase from an annual average of ¥2.5 million in 1879–81 to ¥36.5 million by 1907–9; the annual average for Korea's total trade increased from ¥16.0 million to ¥55.3 million over this same period. The volume of trade continued to expand rapidly during the colonial era, so that by 1939–41 trade had increased more than tenfold in real quantities since 1910–12.[10]

During the colonial era, Korean trade was biased predominantly towards Japan: 85% of her exports went to Japan and 73% of imports came from Japan. As for the composition of trade, Korean exports were mainly raw foods, especially rice, until the late 1930s; rice export accounted for 37% of the value of total exports even as late as 1936–8. However, the relative shares of industrial raw materials and manufactures rose in the 1930s. Imports were dominated by manufactured goods as well as semi-finished manufactured goods, with some low-quality food imports to offset increasing domestic shortages with high-quality Korean rice exports to Japan.

Imports of finished manufactures were mainly consumer goods in the early years, whereas in the 1930s and later they were dominated by machinery and other investment goods needed to expand heavy industry.[11] Much of the new industry needed to support the military build-up was established in the northern part of Korea, because of its geographic and strategic location and the availability of abundant raw materials.

1.4.4 The Second World War

After July 1937, when the Japanese attacked China, Japan had to accelerate its squeeze on the Korean peninsula in order to support its war effort. Japanese colonial government control was extended to encourage production of essential war-oriented goods including foods, to restrict profits, and to curb domestic consumption of food and other materials. In 1943, as a means to stamp out Korean nationalism, Koreans were prohibited from using their own language, and instead had to read only Japanese newspapers, books, and magazines. They were even compelled to change their family names into Japanese. After the attack on Pearl Harbor, Korean people were literally converted into wartime resources for the Japanese: every able-bodied Korean man and woman was forced to work an eleven-hour day and a seven-day week as a contribution towards the war effort.[12] In addition, the Japanese conscripted young Korean men and women as soldiers and as entertaining nurses' aides for soldiers in the front lines and aged Koreans as labourers to work in Japan and elsewhere during 1938–45.

During these periods, the high rate of taxation and rising inflation were used as additional means to fuel the Japanese war machine. For example, the tax rate in 1942 rose to three times that of 1936. Annual consumer price inflation doubled as a result of excessive Japanese currency issue to finance the war and also because of shortages of basic necessities.

1.5 KOREA 1945–53: FROM THE YEAR ZERO TO WAR

> You shall hear of wars and rumours of wars, but see to it that you are not alarmed. Such things must happen, but the end is still to come. Nation will rise against nation, and kingdom against kingdom. There will be famines and earthquakes in various places. All these are the beginning of birth pains.
>
> *Matt. 24: 6, 7, 8*

Japan surrendered unconditionally to the Allied powers on 15 August 1945, but Korean independence took place earlier, on 1 December 1943, when the heads of three of the Allied powers—the United States, Great Britain, and China—signed the Cairo Declaration, which stated: 'The aforesaid three powers, mindful of the enslavement of the people of Korea, are determined that "in due course" Korea shall become free and independent.'[13]

During the 20 months from the Cairo Declaration to the Japanese surrender, the Allied Powers came to no concrete agreement on the nature of their entry into Korea to establish an independent country 'in due course'. The idea of 'in due course' meant some kind of interim international trusteeship over Korea. At the Yalta Conference on 8 February 1945—the meetings of the heads of state of the United States, Great Britain, and the Soviet Union—US President Roosevelt suggested that Korea should be under the trusteeship of the Soviet, American, British, and Chinese governments. He said: 'the only experience we had had in this matter was in the Philippines where it had taken about fifty years for the people to be prepared for self-government'.[14] He said that 'in the case of Korea the period might be from twenty to thirty years', but Stalin replied, 'the shorter the period the better'.[15] Marshal Josef V. Stalin also enquired as to whether any foreign troops would be stationed in Korea. The President replied in the negative, to which Stalin expressed approval.[16]

In the six months between the Yalta conference and Japan's surrender on 15 August 1945, several very important international events occurred. In April President Roosevelt died and Harry Truman succeeded to the role of US president; in May Germany surrendered; in July the Allies met again at Potsdam; on 6 August the United States dropped the first atomic bomb on Hiroshima.[17] Two days later the Soviet Union declared war on Japan, exactly as Stalin had told Harry Hopkins in May and had promised in the Potsdam Declaration.[18] By that time, the Russian interest in Korea was more explicit than that of America; since the Korean peninsula is close to the Soviet Union it is important to Russian security and foreign trade. Soviet troops entered northern Korea on 10 August and successfully set up a bridgehead from which to play a forceful role in post-war north-eastern Asia. This Soviet entry, or the immediate prospect thereof, triggered the American fear that the peninsula might fall wholly into Soviet hands unless US troops made their own entry.[19] On 8 September American troops landed on the Korean peninsula from Okinawa, 600 miles away, under the command of Lt-General John R. Hodge. Consequently, the two superpowers agreed to divide Korea into areas north and south of the 38th Parallel for the purpose of military operations of accepting the surrender of Japanese troops. The border was a temporary expedient, and nobody expected at that time that the line would become a friction spot of power politics, and destroy Korean national unity. Unfortunately for the Korean people, the line of temporary tactical considerations turned out eventually to divide the country into Russian-occupied North Korea and American-occupied South Korea. The division extends through all phases of Korean life—economic, political, and social. The division became another watershed in Korean history that led the two halves of the country along different paths of destinies of opposing systems, resulting in a fratricidal war and many other human tragedies.

Of course, during the period between 1945 and 1948, there had occurred several meetings among the Allied powers to impose trusteeship over Korea. In December 1945 the powers agreed to form a joint Soviet–US commission to assist in organizing a single 'provisional Korean democratic government'. The trusteeship proposal was immediately opposed by all Koreans except the communists, who also objected at first but under Soviet pressure quickly changed their position.

The joint commission met in Seoul several times between March

1945 and October 1947. The Soviet side insisted that only those democratic parties and social organizations upholding the trusteeship plan should be allowed to participate in the formation of an all-Korean government. But the nationalist groups, backed by the United States, rejected the Soviet formula, arguing that once it was accepted Korea would be put under the hands of communists.

In September 1947, the United States submitted the Korean question to the General Assembly of the United Nations, which in November, over Soviet objections, adopted a resolution stipulating that the elected representatives of the Korean people choose their own form of government. The UN Temporary Commission on Korea was organized to observe nationwide free elections. After the Soviet refusal in January 1948 to admit the commission to the northern half, elections were held on 10 May 1948 in the southern half only. Meanwhile, the North Korean communist leaders prepared a constitution of their own and adopted it in April 1948, and it was ratified officially on 3 September by the Supreme People's Assembly for Korea. The Democratic People's Republic of Korea (DPRK) was officially established on 9 September 1948, and Kim Il-Sung was elected its president in the North as the counterpart to President Syngman Rhee of the Republic of Korea (ROK) government in the South, which was set up as the result of the UN-supervised election.

At the start of the transition period when the country was being divided (1945–8), the southern part of Korea encompassed primarily the agricultural area. It had an area of 98 430 km^2 with a population of 9.3 million. Geographically, the southern zone was blessed with relatively rich soil and a mild climate, thus having a comparative advantage in agricultural activities, particularly rice production, while the northern part was generally endowed with industrial prerequisites, having rich mineral resources and hydroelectric power supply from the Yellow River power generation station built during the Japanese colonial era. Before the division, northern Korea had 75% of the heavy industry, while southern Korea had almost 75% of the light industry. Southern Korea had almost three times as large an area of irrigated rice paddies as the north. The south depended on the chemical fertilizer produced in the north as well as the north's electrical power supply, while the north needed rice from the south. The north shut off power transmission to the south when UN-supervised elections were held in

TABLE 1.4 Casualties of the Korean War, 1950–1953

	Civilian		Soldiers		Others	
	North Korea	South Korea	North Korea	South Korea	Chinese	UN soldiers
Deaths	406 000	373 599	294 151	227 748	184 128	36 813
Wounded	1 594 000	229 652	225 849	717 083	715 872	114 816
Missing	680 000[a]	387 744[b]	91 206[c]	43 572	21 836[c]	6 198
Total	2 680 000	990 995	661 206	988 403	921 836	157 827

[a] Refugees into the South are included.
[b] Of which, 85 532 persons were taken to the North.
[c] Reported prisoners of war.

Source: Young-Hwan Kil, *Comparative Politics between North and South Korea* (in Korean) (Seoul: Moon Mack-Sa, 1988), 65.

the south on 10 May 1948, thus completing the severing of north–south ties. The traditional complementarity between southern part and northern part was suspended as two governments emerged synchronically in north and south with antagonistic eyes on one another.

Just after the communist-controlled government was set up in the north, Soviet troops withdrew from the area. About six months later, in mid-1949, the US forces went home from the South. That occurred a few months before US Secretary of State Dean Acheson outlined an American retrenchment in Asia with a Pacific 'defence perimeter' that did not include Korea. But on 25 June 1950 the Korean war broke out. This resulted in three years of bloody conflict and a stalemate on the battlefield with intervention by the United States and later China.[20] Under a 1953 truce agreement at Panmunjom, a demilitarized zone was established near the 38th Parallel. Since then, the Korean peninsula remains one of the major unresolved battlefields of military adversaries between North and South and of the Cold War between East and West.

North and South Korea have since followed different paths of economic development against a background of a common language, cultural history, and sense of identity. Opposing political ideologies, along with their corresponding economic systems, have created antagonistic relationships and feelings of bitter enmity between the two halves of the country. Particularly since the Korean War (1950–3), the two Koreas seem to have kept themselves remote from each other across the 155 miles of truce line which remains as a border of 'no return'.

NOTES TO CHAPTER 1

1. Korean Overseas Information Service, *A Handbook of Korea* (Seoul, 1988). Also see Lee Ki-Paek, *Hankuk-sa Sinron* (Seoul: Seoul International Publishing House, 1962).
2. Chong-Sik Lee, 'Historical Setting', in W. Evans-Smith (ed.), *North Korea* (Washington, DC: American University, 1981), 11–12.
3. Bong-Youn Choy, *A History of the Korean Reunification Movement: Its Issue and Prospects* (Peoria, Ill.: Bradley University, 1984), 7–8.
4. Donald G. Tewksbury, *Source Materials on Korean Politics and Ideologies* (New York: Institute of Pacific Relations, 1950), 21–3; C. I. Eugene Kim and Han-Kyo Kim, *Korea and the Politics of Imperialism, 1876–1910* (Berkeley and Los Angeles: University of California Press, 1967), 126.
5. Bong-Youn Choy, *Korea: A History* (Rutland, Vt., and Tokyo: Charles E. Tuttle, 1983), 118.
6. Korea Overseas Information Service, *Handbook of Korea*, 94–5.
7. Paul W. Kuznets, *Economic Growth and Structure in the Republic of Korea* (New Haven, Conn.: Yale University Press, 1977), 13–14. See also Sang-Chul Suh, 'Growth and Structural Changes in the Korean Economy Since 1910', Ph.D. dissertation (Harvard University, 1966).
8. Ibid. 22. See also Choi Ho-Chin, 'The Process of Industrial Modernization in Korea: The Latter Part of the Chosen Dynasty through 1960s', *Journal of Social Sciences and Humanities* (Seoul), 26 (June 1967): 1–33.
9. Andrew J. Grajdanzev, *Modern Korea* (New York: John Day, 1944). See also Kuznets, *Economic Growth*, 22–3.
10. Su, 'Growth and Structural Changes', B54–B56 and tables II–11 and II–14; 'Distribution of Trade Ratios', *Bank of Chosen Economic Review*, 3 (1949), 426–34 and table 56. See also Kuznets, *Economic Growth*, 9–10, and Hilary Conroy, *The Japanese Seizure of Korea: 1868–1910* (Philadelphia: University of Pennsylvania Press, 1960), 456–7.
11. Kuznets, *Economic Growth*, 11–13.
12. Choy, *Korea*, 31–2.
13. US Department of State, 'Foreign Relations of the United States: Diplomatic Papers, Conference at Cairo and Teheran, 1943' (Washington, DC: US Government Printing Office, 1961), 399–404.
14. George M. McCune, *Korea Today* (Cambridge, Mass.: Harvard University Press, 1950), 103.
15. Robert E. Sherwood, *Roosevelt and Hopkins: An Intimate History* (New York: Council on Foreign Relations, 1956), 11–12.
16. US Dept of State, 'Foreign Relations', 1770.
17. Jong-Chun Baek, *Problem for Korean Unification* (Seoul: Research Centre for Peace and Unification, 1988), 32–43.
18. US Department of State, *The Record on Korean Unification, 1943–1960* (Washington, DC: US Government Printing Office, 1960), 43–4.

19. Gregory Henderson, 'Korea', in G. Henderson *et al.* (eds.), *Divided Nations in a Divided World* (New York: David McKay, 1974), 43–96.
20. The Korean War began on 25 June 1950 and ended with a military armistice (rather than a peace treaty) on 27 July 1953. During these three years, the number of casualties was estimated as shown in Table 1.4.

2

The Evolution of Two Rival Systems

Let them grow side by side until harvest time, and at harvest
time I shall direct the reapers to collect the weeds first, bundle
them up and burn them, but bring the grain into my barn.

Matt. 13: 30

2.1 GENERAL BACKGROUND

With the defeat of Japan at the end of the Second World War,
Korea was liberated from the Japanese yoke. While the Korean
people were caught up in a celebration of the nation's independence, however, destiny worked to divide the peninsula in half.
There followed a bloody conflict (1950–3) that was suspended by
an armistice but never concluded. Since then, South Korea has
followed the capitalist road of a market system as introduced by
the United States with huge amounts of subsequent economic aid
intended to help it recover from the depths of economic misery.
North Korea, in contrast, has adopted and practised the Marxist–
Leninist command economy as imposed by the Soviet Union.

The fundamental elements of the market economic system on
which the South Korean economy is based are characterized as
(1) privatization of the means of production (capital, labour, and
natural resources); (2) the diversification of decision-making processes; and (3) the built-in stabilization mechanism, which operates
principally in accordance with market laws. The general merit of
a market-oriented economic system can be found in its efficient
allocation of resources, advancement of technology, and the realization of consumer sovereignty and diversification of social
powers which accompanies individual freedom and responsibility.
However, often the capitalist market not only produces so-called
market failures such as imperfect competition, externality, and an
insufficient supply of public goods, but also can result in economic

instability arising from depression, unemployment, and inflation, which can lead to a growing inequality of income and wealth distribution. These problems inevitably necessitate the partial intervention of government into markets by means of indicative macroeconomic planning or adjustment policies. Such policies include progressive income tax systems and social welfare programmes aimed at promoting horizontal and vertical equity.

The South Korean economy has been no exception. It has adopted overall economic development plans to speed up economic growth and development since 1962. These plans are basically in line with a free-market system and are intended to provide broad guidelines for the macroeconomy to move forward. In a market economy, however, it is not always evident what planning does or whether it plays a significant role in the economy. Indeed, empirically it is not possible to evaluate the effects of government planning on the economy compared with the performance of that economy in the absence of such government planning. The plans reveal what has been done to promote development, and raise the issue of the government's proper role in a free-enterprise economy —in particular, whether economic performance would improve if the government did less and markets moved to allocate resources.

The South Korean experience suggests that planning has increased and improved the information needed to make economic decisions and has co-ordinated government economic policies. The government has aided the development of new industries and raised export competitiveness, by providing direct assistance and incentives to private entrepreneurs to implement planning goals. However, the 1980s saw a decline in the indicative, co-ordinative, and prescriptive functions of government activity in South Korea. As the economy has become increasingly complex, and as the country's liberalization and democratization processes complicate the planner's objective function, the decline in planning is quite natural. In a sense, it is a product of the changing environment, in terms of the cost–benefit ratio in planning, as the scale of the economy expands. In other words, it reflects the growing understanding of the importance of incentives and markets *vis-à-vis* the observed inefficiencies of planning as the economy moves into a more highly developed phase.

North Korea's economy, on the other hand, can be said to be more centralized, more controlled, and more ideologically orthodox and monocratic than those of any of the world's other communist states. With its autarkic command and rationing system, the North Korean economy is understood to have put great emphasis on social equity and welfare.

The North's economic policy regimes must have been grounded principally on Karl Marx's hypothesis that the structure of social relations in the economic subsystem (and their changes) exerts a more powerful pressure upon the subsystems of law, opinion, politics, and ideology.[1] But in retrospect, it seems to have worked the other way around, because President Kim Il-Sung's ideology has underlain all North Korean economic policies and directions, as well as most other areas of everyday life.

North Korea manages its economy according to the ideology of *chuche* (self-reliance; independence), on which politics and economics are close supplements of one another. The idea of *chuche* was first introduced by Kim Il-Sung at North Korea's Workers' Party Functionary Meeting on 28 December 1955.[2] This was a self-protective fact adopted by Kim's regime for independence and self-reliance from foreign dependency in the wake of the ideological dispute between the Soviet Union and China after the death of Josef Stalin in 1953.

The '*chuche* ideas' were called upon for various political and economic reasons or purposes. First, the '*chuche* for ideology' was introduced by Kim Il-Sung to the Workers' Party Central Committee in 1955, as mentioned above. The basic idea behind this *chuche* was to make all North Koreans want to become 'ardent communists' for revolutionary tasks. Next, the '*chuche* for economy' was proposed by Kim on 11 December 1956. At that time, North Korea was experiencing difficulties in securing continuing economic aid from both the Soviet Union and China. This '*chuche* idea' calls for a disproportionate emphasis on the development of an independent, self-reliant economy which can display the superiority of the socialist system of North Korea. It calls on working people to be the masters of plants and farms, and also of facilities and materials. Thirdly, the '*chuche* for self-defence' was introduced at the general meeting of the Workers' Party Central Committee on 10 December 1962.[3] Lastly, the '*chuche* for political independence' appeared in the editorial of the *Rodong Sinmun* (the government newspaper in North Korea) on 12 August 1966, quoting Kim Il-Sung's remarks on that issue.

Thus, the '*chuche* idea' evolved and grew in importance from its birth in 1955 to 1966, when it was praised for being a 'unique thought' comparable to the 'Marxist–Leninist line'. At the fifth Workers' Party meeting in November 1970, the '*chuche* idea' was officially recognized as the 'party's guiding ideology' along with the 'Marxist–Leninist philosophy'. Two years later, it was designated as the 'leading guideline' for the country in the new 'communist constitution' promulgated in December 1972.

Vigorously conducting *chuche* ideal indoctrination by linking it with the implementation of the revolutionary tasks, the North Korean élites mobilize the potential of the masses for increasing production and conservation. The party has recently been stressing the tasks to increase the production of electricity, coal, and steel, to solve the transport problem, and to fulfil the agricultural production target by completing irrigation and chemicalization programme in accordance with the demands of the *chuche*-oriented agricultural method.[4]

The '*chuche* idea' was further developed and enriched by Kim Chong-Il, son and heir-designate of Kim Il-Sung, as a guiding force of the 'Three Revolutions'—ideological, technical, and cultural.[5] It is apparent that the '*chuche* idea' is now being used as a tool for strengthening the political position of Kim Chong-Il in the power transition process that is under way in the North. As in other communist economies, the means of production are owned completely by the people in the form of 'all-people ownership' and co-operative ownership, which in theory is geared to serve the promotion of the material well-being of the masses.[6]

One other major difference between a market economy and the socialist system lies in the decision-making criterion as to 'what', 'how', and 'for whom' to produce. In deciding *what* to produce, consumer demand plays a very important role in the market-oriented economy, while social preference ordering by value judgement of the ruling leaders dominates in the command economy. In the latter case, therefore, the priorities of production are ranked down from public goods, producers' goods, light industry goods, and lastly consumer goods, so that consumer preferences on the micro level are neglected for the sake of the macro-level targets of the planners. The emphasis on heavy industrial and military sectors has become doctrine in most socialist economies since the Stalinist period, and North Korea was no exception.[7] A capitalist economy traditionally puts the emphasis on the development of consumer goods and light industry in the early stages of economic development; then, as it accumulates capital over time as the economy grows, it expands into heavy industry in later stages. The socialist industrialization method is to develop heavy industry from the beginning. Therefore, it needs to secure a large amount of start-up capital. North Korea attempted to mobilize capital requirements not from foreign savings but rather from the so-called 'socialistic accumulation'[8] in line with its independent/self-reliant doctrine. However, the sources for expanding social resources appear to have been severely limited, apart from urging the people to tighten their belts and encouraging productivity increases,[9] since the North's net income growth has been mediocre.

The decision on *how* to produce also provides good comparisons of the two systems. The capitalist system pursues the maximization of profits, thus assuming efficient allocation of resources through market signals unless its institution and information flows are imperfect. In a communist economy, the means of production are socially owned and therefore the role of major players—state corporations and co-operative enterprises—are strong. The activities of state and co-operative firms are mostly motivated towards political and social objectives. Allowable economic actions are in the hands of the top planning authority which behaves in accordance with given resource constraints. Thus, the market is largely irrelevant in production decision-making. The central plan specifies a large number of value aggregates and physical inputs and outputs in the economy; in this system money plays only a passive, accounting role. The structure of the plan is strictly hierarchical, so that the lower levels are formally subordinated to those above. The plans are enforced by rationing the means of production—materials, labour, capital—rationing of goods and services, and the administrative allocation of manpower and job targets. By and large, competition is limited to efforts at plan fulfilment and over-fulfilment, except for a small (legal, semi-legal, and illegal) private market for some farm produce and services.

As to the distribution of outputs, the two systems differ in terms of the choice of beneficiary or demander. The capitalist system distributes its products through a market auctioning mechanism to those who can afford to pay the asking price. In other words, output flows in the direction of meeting demand. Often consumers face income and wealth constraints which result from the distribution curves of initial wealth endowments, the abilities of each individual, and other social factors.

The markets are mostly buyers' markets in a capitalist system except for a number of natural monopolistic goods. Therefore, the producers and managers must compete to provide the prospective buyers with relatively less expensive but better-quality goods and services. In the communist system, on the other hand, the top decision-makers ration outputs to the needy in accordance with their needs. Usually priorities of output distribution go to defence, education, medical care, and consumption, in that order. The share allocated for final household consumption is always marginal so long as the economy is still supply-constrained.

The North Korean planning authority, like those in other communist economies, has developed its own material incentive systems which are coupled with basic wages, bonuses, and awards of medals as means of encouraging the fulfilment or overfulfilment of obligatory plan targets, usually defined in percentage growth rates of physical quantities.

In North Korea, the basic wage per every 1% fulfilment rate of plan target is $V/100$, where V is the plan's basic straight wage rate. The reward (bonus) scale if fulfilment exceeds the target rate is based on the following formula:

$$\frac{\text{Plan's anticipated rate of total product value } (C + V + M)}{100},$$

where C is material cost, V wages, and M net material income.[10] Note that $V + M$ represents net value of social product in the socialist economy, and the rate of state plan fulfilment is basically targeted to be 100%. The reward (bonus) per every 1% overfulfilment exceeding the state plan target (100%) will grow larger than the basic wage set for plan target accomplishment by a factor of $[100/V/(C+V+M)]$. For example, if we assume that the wage rate accounts for 20% of a unit product price, then the size of reward (bonus) for every 1% overfulfilment will be five times (that is, 100/20/100) greater than the growth rate of basic wage achieved within the limits of plan goal.[11] From this, it can be deduced that the trade-offs between labour productivity growth and wages and bonuses are crucial for the socialist accumulation. To increase communist accumulation, of course, the wage and bonus growth rate must be kept equal or less than the increase in labour productivity, if other conditions are unchanged.

In retrospect, the reward system linked to the productivity of workers and managers in a controlled economy may not succeed in continuously bringing higher growth rates. Workers may be cautious about keeping up the overfulfilment of the plan by too large a margin. They know that production levels in one period will be the government basis for setting targets in the next period. Unless the reward and bonus are large enough to compensate for their increased toil, workers will find little reason for maintaining their efforts, unless persuaded by blind patriotism or fidelity to their leaders. Sooner or later, the communist command economy is likely to face a pitfall where 'carrot' (reward or bonus) or 'stick' (mobilization drive) can no longer be used, in spite of the needs for higher communist accumulation and economic growth.

North Korea appears to be in need of reforming its economy, as are all other communist countries, in an attempt to revive the market. It is now the only one of the Soviet bloc countries that shows no sign of major changes in its tightly closed economic system and policy.

2.2   TRANSITIONAL REFORMS, ECONOMIC PLANS, AND MANAGEMENT SYSTEMS

2.2.1   *North Korea*

Reforms in transition

Under the slogan of 'communist economic construction', North Korea has aimed at building a self-reliant economy under a central planning and management system. As briefly discussed in Section 2.1, its priorities for economic development were on heavy and military industries rather than consumer goods and agriculture. The strategy of 'economic self-reliance' is synonymous with the inward-oriented economics of an autarkic state.

North Korea has become an orthodox communist state of its own type, although its founding philosophy was imported from the Soviet Union. It has followed the 'Marxist–Leninist line', complemented later by Kim Il-Sung's '*chuche* idea', with fidelity and enthusiasm in the field of economic planning and management as laid out in both the early Five-Year Plans of the Soviet Union and in the similar pattern of communism in China. Philip Rudolph explained this point as follows:[12]

In overwhelming measure North Korean political and economic institutional development has been patterned on that of the Soviet Union ... By utilizing Soviet experience, North Korea could repeat much of the Soviet development pattern in a much shorter time ... Such differences as exist between the Soviet and North Korean economic institutions are in the realm of theory rather than practice (as in the case of the Soviet collective farm and the North Korean cooperative). North Korea's pattern of socialization has in many respects resembled that of China rather than that of Eastern Europe, particularly in economic policy since the end of the Korean War. The timing and tempo of collectivization (although different from the Chinese commune system) closely coincided with that of China ...

During the transitional and post-war years spent in building the People's Democracy, had the autonomy of the North Korean administration been kept intact even under the Sino-Soviet models? The evidence is that Kim Il-Sung did manage to exercise autonomy by keeping an equal distance between the Soviet Union and China.[13]

Between March and August 1946, the North Korean Interim People's Committee (NKIPC), under the guidance of Russian Advisory Group, enacted a series of radical reforms covering every field of policy: agrarian reform (5 March), labour law (24 June), law on equality of the sexes (30 July), and law on the nationalization of industry, transport, communications, banks, etc. (10 August).[14] All these reforms bore Korean characteristics, although the ideas for the reforms were borrowed from the Soviet Union under the 'general principle' of communist revolution.

The 1946 land reform, for example, was probably more successful than is generally recognized in the West. It was a basic 'land-to-the-tiller' type of reform, establishing smallholdings, not collectivized farms. It had two striking features. First, it was by far the fastest, and probably the most peaceful, land reform in any communist country ever, officially taking only three weeks to implement. Second, it allowed large landlords to own the same size farms as anyone else, provided that they moved to another district. (Most in fact fled to the South.) Of course, this land reform was changed with the introduction of 'co-operative farms' in the mid-1950s.[15]

Economic plans: goals and performances

> An economic life constitutes the basis of social life. The so-
> cialist economic life must be sound, contributing to the ful-
> filment of the people's desire for independence and to the
> provision of their creative activities; it must be equitable for
> everybody, providing him with equal happiness without gaps
> between the rich and the poor.

> *President Kim Il-Sung's policy speech to the first session of*
> *the Ninth Supreme People's Assembly of the DPRK at the*
> *Mansudae Assembly Hall, 24 May 1990*

North Korea completed the socialization of productive relationships by enacting a series of reforms covering land (March 1946),
the nationalization of major industries (August 1946) and of forestry, water basins, and mineral resources (December 1946), and
the formation of farmers' co-operatives (consisting of about 300
households each), handicraftsmen, merchants, and industrialists
(August 1958). Through this socialization process, the major means
of production were brought under centralized control. Most of the
land not under the operation of the farm and fishery co-operatives
belongs to the state, as do other productive resources such as
natural (mineral) resources, forests, harbours, major factories and
other enterprises, banks, and transportation and communication
facilities. The co-operatives own their own land, farm equipment and machinery, animals, ships, buildings, and some factories
and business facilities. Private ownership is permitted for non-
productive resources subject to private consumption, however. For
example, small plots of farm land (less than about 260 m^2), allocated
to individual farms for the production of foodstuffs, either for
their own use or for sale at open markets, were classified as being
in private ownership. In 1977 the state reduced the authorized size
of private plots from 260 m^2 to 66–99 m^2.[16]

The economic policy regimes of North Korea can be chrono-
logically classified as follows.

(a) Liquidation period: 1945–1946 In laying the foundations for
a People's Democracy, in a country where exploiting elements still
existed, the basic aims were to take over Japanese colonial industrial facilities, reconstruct the war-damaged economy, and put it in
good order through nationalization. To this end, the government

instituted agrarian reforms and nationalized major industries during this period. Table 2.1 shows the process of land reform (land confiscated from and land reallocated to) conducted in North Korea during this period.

(b) The two One-Year Plans of 1947 and 1948   Major economic policy goals were targeted (1) to consolidate factories and enterprises in workable order, (2) to promote the production of necessary goods, and (3) to encourage agricultural production. The value of total industrial output grew by 54% in 1947 and 64% in 1948, of which producer goods rose by 76% and 78% while consumer goods increased by 30% and 50% in the respective years. The output of grains (unpolished) was 1 898 000 tons in 1946, 2 069 000 tons in 1947, and 2 668 000 tons in 1948, recording an annual average growth rate of 11.9% during 1947 and 1948.[17]

(c) The first Two-Year Plan: 1949–1950   North Korea's government launched its first Two-Year Plan with a major emphasis on consolidating the foundations for a self-reliant national economy by seeking technically to improve both the manufacturing and agricultural sectors. Official statistics reported for 1949, compared with those for 1946, show that national income increased 210% and total industrial output 340%. The growth rate of total industrial output was 119% in 1949 over 1948, of which producers' goods increased by 121% and consumer goods by 108%. Producer goods comprised 62.0% of total industrial output, while consumer goods accounted for only 38.0% in 1949, compared with 66.5% and 33.5%, respectively, in 1946.

The value of agricultural products rose by 51.1% from 1946 to 1949. Meanwhile, the number of employees increased by 117%, and factory and office workers' salaries increased by 83% over the same period. The overall productivity of all industrial workers increased by 250%.[18]

This plan was interrupted by the Korean War, which lasted from June 1950 to August 1953.

(d) The post-Korean-War Three-Year Plan: 1954–1956   To restore the War-devastated economy, North Korea made great effort with the economic aid received from communist bloc countries, mainly China and the Soviet Union.[19] Because of the emphasis on 'vast

TABLE 2.1 Farmland reform in North Korea, 1945–1946

	Area (cheongbo[a])	Cultivated crop land (cheongbo[a])	Orchards (cheongbo[a])	No. of families affected
Land confiscated: total from	1 000 325	983 954	2692	422 646
Japanese	112 623	111 561	900	12 919
National traitors and fugitives	13 272	12 518	127	1 366
Holders of more than 5 cheongbo	237 746	231 716	984	29 683
Nonfarm-owners	263 436	259 150	292	145 688
Landowners renting out	358 053	354 093	381	228 866
Religious groups	15 195	14 916	8	4 124
Land allocated: total to	981 390	965 069	—	724 522
Hired farmers	22 387	21 960	—	17 137
Farmers with no land	603 407	589 377	—	442 973
Farmers with little land	345 974	344 134	—	260 501
Emigrating landowners	9 622	9 598	—	3 911
People's committees[b]	18 935	18 885	2692	—

[a] 100 cheongbo = 99.1739 hectares.
[b] Not included in land allocated to farmers.

Source: People's Democratic Republic of Korea (DPRK), *Statistics of the People's Economic Development, 1946–1960* (in Korean) (Pyongyang: National Press, 1961), 59, table 36.

capital construction', the state spent W80 582 million[20] (about US$322 328 000), which was W1800 million more than originally planned. Out of the total investment during the period of the Three-Year Plan, W58 938 million (73.1%) is said to have been made available for 'productive construction',[21] of which W39 948 million (49.6%) was allocated for industrial construction, W7443 million (9.2%) for agriculture, and W10 564 million (13.1%) for transportation and communication. The remaining W21 644 million (26.9%) went towards educational, cultural, and other establishments, of which W9683 million (12%) was used for private dwellings.

Table 2.2 shows some basic statistics of the North Korean economy in the 1950s. The Three-Year Plan (1954–6) was claimed to have been fulfilled in two years and eight months, which enabled North Korea to demonstrate the superiority of its communist system over the then poverty-stricken economy of the South.

It must be noted that the first year of this Three-Year Plan period was the transitional point when the then private farmers were told by the Workers' Party that they were expected to participate in the Co-operative Movement. In August 1953, there were only 174 co-operatives, but the number swelled to 74 000 co-operatives in November 1954, of which 219 000 farm households were members. The farm co-operatization drive was complete by August 1958. By that time, the average size of each farm co-operative encompassed about 80 farm families with total cultivated land size of 130 cheongbo (1.29 million square metres or 318.6 acres). The size of each co-operative in this initial stage was relatively small, because they were organized long before the farm mechanization programme was completed.[22]

(e) The first Five-Year Plan: 1957–1961   The objective of this plan was both to strengthen the material base for a communist economy and to ensure a supply of basic necessities for the people. During the period of this plan, the socialist transformation of most production relationships was completed, making it possible for the state to control all production activities in the country. For example, the proportion of outputs under the control of state enterprises and co-operative groups in the manufacturing and mining sectors expanded from 72.4% in 1946 to 98.0% in 1956 and then to 99.9% in 1958. The farm co-operative share in agriculture rose from 3.2% in 1949 to 73.9% in 1956, to 88.2% in 1957, and to 100.0% in 1958.

TABLE 2.2 Selected indicators of the North Korean economy, 1946–1960

	1946	1947	1948	1949	1950	1951
Social value of products (%)	100			219		
National income (%)	100			209		
Industrial output (%)	100	154	213	337	295	157
Producer goods (%)	100	176	254	375	333	123
Consumer goods (%)	100	130	180	288	254	187
Agricultural output (%)	100			151		
Cultivated land ('000 cheongbo)[a]	1860			1983		
Paddy	388			467		
Upland	1472			1516		
Investment (W million)[b]	66					
Industry						
Heavy						
Light						
Agriculture						
Transport and communication						
Commerce and social supply						
Education and culture						
Research						
Health						
Housing						
Public facilities						
Other						
Consumer price index (%)				100		
Total population[c] ('000)	9257			9622		
No. of employees ('000)	260.0	367.6	442.6	565.0	486.0	351.8
Labour productivity growth[f] (%)	100			252		
Wage growth (%)				100		

[a] 0.40806 cheongbo = 4046.8 m^2 = 1 acre. Therefore 1 cheongbo = 9917.2 m^2 = 2.4506 acres.
[b] W million of North Korea in 1950 constant prices.
[c] At end of year.
[d] At 1 December of the corresponding year.
[e] At 1 September of the corresponding year.
[f] Labour productivity growth rate in manufacturing sector.

1952	1953	1954	1955	1956	1957	1958	1959	1960
	163			355			735	797
	145			319	417	594	636	683
178	216	326	485	616	890	1 218		2 100
136	158	299	488	640	936	1 262		
218	285	366	497	598	878	1 190		
	115			160	198	252		224
	1 965			1 899	1 907	1 924		1 913
	478			491	502	508		510
	1 487			1 408	1 405	1 416		1 403
		24 831	29 349	26 402	27 136	34 122		
		10 729	15 075	14 144	15 701	18 619		
		8 686	11 933	11 778	13 183	15 828		
		2 043	3 142	2 366	2 518	2 791		
		1 584	3 092	2 767	1 395	2 980		
		4 956	3 521	2 087	1 708	2 637		
		309	276	399	858	1 154		
		2 055	1 431	1 278	1 259	1 194		
		67	145	126	153	194		
		329	497	315	430	506		
		2 625	3 201	3 357	4 060	4 373		
		1 944	1 540	1 133	1 112	2 060		
		233	571	296	460	405		
	265	197	182	165	159	156		135
	8 491[d]				9 359[e]		10 392[d]	10 789
418.3	574.6	690.0	763.3	808.2	844.5	983.0		
	197			386	477	506		539
	105	127	141	165	236	256	365	386

TABLE 2.3 Types of ownership in North Korea, selective years, 1946–1960 (%)

	Socialist ownership			Small commodity dealers' ownership (retailers)	Capitalist ownership (private)
	Total	State	Co-operatives		
1946	19.1	18.9	0.2	60.9	20.0
1949	47.6	43.7	3.9	44.2	8.2
1953	50.5	45.1	5.4	46.6	2.9
1956	89.0	60.2	28.8	8.7	2.3
1959	100.0	68.1	31.9		
1960	100.0	69.1	30.9		

Sources: Chosen Chungang Nyungam (1961), 323, 334; DPRK *Statistics Book, 1946–1960* (Pyongyang, 1951), 23, table 8.

In the retail and commercial sectors, the share of outputs in the hands of both state and co-operatives accounted for 47.6% in 1949, but it jumped to 89.0% in 1956 and to 100% in 1959.[23] By 1959, therefore, most individual retailers had disappeared from the markets (see Table 2.3).

North Korea reported that, during this period of the first Five-Year Plan, the total value of industrial output had risen more than threefold as compared with the plan's target value of 2.6 times, and the total value of commodities in circulation had increased by about 2.5 times.[24] It has been claimed that the Five-Year Plan was completed within two and a half years, the rest being called a 'transitional period' for the preparation of the second Five-Year Plan, which would begin in 1961 instead of 1962 as originally planned.[25] This is now known as the Seven-Year Plan.

Table 2.4 shows annual average growth rates of the value of industrial outputs and trends of investment composition between heavy and light industries during the 1947–60 period. Table 2.5 gives major industrial growth patterns over the same period. From this table, it can be seen that North Korea achieved great progress, particularly in machinery and metal industry, textiles, construction materials, hide, and rubber industries.

TABLE 2.4 Growth rate of outputs and composition of industrial
sector investment, North Korea, 1947–1960 (%)

	1947–9	1954–6	1957–60
Value of industrial output	49.9	41.7	40.3
Producer goods sector	53.3	59.4	42.1
Consumer goods sector	42.3	28.0	38.6
Investment share of industrial sector[a]	40.9	49.6	51.3
Heavy industry	86.0	81.1	82.6
Light industry	14.0	18.9	17.4
Investment share of agricultural sector		9.2	10.5

[a] Share of industrial sector investment out of total investment expenditure, in current
prices.

Source: *Chosen Chungang Nyungam*, 1949–1962.

It is noted in passing that North Korea devised two major
labour mobilization schemes during the shortened plan period
of 1957–60. One is called 'Cholli-ma Wundong' ('Flying Horse
Movement'), which was initiated in 1958 to systematically mobilize labour forces and thereby speed up the development of heavy
industry.[26] The other one is the 'Chongsan-ri Băng-bŭp' ('Blue
Mountain Village Method'), which started first with President Kim
Il-Sung's 'on-the-spot' guidance in February 1960, when he visited
the Chongsan-ri co-operative farm in Kang-sĕo county, Pyongnam
province. It is also notable that North Korea implemented its
second currency reform in February 1959, by which 100 old *won*
was denominated to 1 new *won*.[27]

**(f) The first Seven-Year Plan (1961–1967) and its three-year
extension: 1968–1970** The first Seven-Year Plan (1961–7) of North
Korea was actually the first intensive economic development plan
of its kind. North Korean officials focused on an inward-oriented
industrialization policy with the emphasis on heavy industry, as
had evolved under Brezhnev in the Soviet Union. Kim Il-Sung
adhered to the view that the route to an advanced, affluent communist economy was through machine tools and other heavy industry. It is not hard to see that North Korea's focus on machine
tools was connected with its desire to upgrade its military sector.

TABLE 2.5 Growth of major industries in North Korea, 1946–1960 (%)

	1946	1949	1956	1959	1960
Electricity generation	100	151	130	203	234
Fuel industry	100	301	240	563	659
Mineral ore mining industry	100	398	561	1 200	1 400
Metallurgical industry	100	388	567	1 300	1 700
Machine and metal processing	100	535	2 100	8 300	9 900
Chemical industry	100	310	288	1 100	1 300
Wood processing and timber	100	173	307	579	566
Textile industry	100	685	2 100	6 100	7 200
Cultural and daily goods industry	100	616	2 200	10 600	15 000
Fishery industry	100	779	720	1 600	7 200
Food and luxury goods industry	100	238	301	1 200	1 300
Construction materials industry	100	871	3 100	13 500	14 900
Glass and ceramic industry	100	166	944	4 500	5 000
Pulp and paper industry	100	228	486	1 600	2 000
Printing and publication industry	100	207	674	2 100	2 200
Hide and leather industry	100	2 000	6 400	13 000	15 500
Rubber industry	100	3 000	3 300	8 800	10 100
Oils and fats industry	100	215	1 100	1 200	1 200

Source: *Chosen Chungang Nyungam* (each year).

But it is difficult to explain the planners' obsession with an inward-oriented heavy industry policy as opposed to a focus on exports and consumer goods. The best explanation seems to be that, as Sino-Soviet relations worsened in 1961–2, the North Korean desire to maintain a neutrality between the two communist giants reduced it to pursuing an inward-oriented and self-reliant economic policy. The demagogic '*chuche* idea' was hyped at this stage in the ideological conflict between the Soviet Union and China. As Kim Il-Sung faced a serious problem with respect to Soviet bloc relations, he sought to promote his '*chuche* idea' to every corner of life.

The first Seven-Year Plan was unrealistically weighted towards heavy industry. The heavy industry sector was viewed to have 'big-push effects' through both forward and backward linkages. Development of 'Sector A' industries, as the heavy industry sector was called, was regarded as a necessary prerequisite for the growth of both 'Sector B' industries (consumer goods) and 'Sector C' industries (intermediate goods and inputs).

Along with this heavy-industry-oriented policy, North Korean policy-makers also emphasized the importance of revitalizing a grass-roots agricultural sector through the Chŏngsan-ri spirit and the *chuche*-oriented agricultural method,[28] but the achievement in the agricultural sector during the plan period was not realized.

The original Seven-Year Plan was extended to accommodate three more years to establish a combined Ten-Year Plan (1961–70). The main tasks of this plan were to raise national income 2.7 times, industrial output 3.2 times, and grain production 6 million to 7 million tons over 1960. North Korea asserted that it had increased industrial output 330% and machine tool industry 220%, along with an overall labour productivity growth of 147.5%, during the 1961–70 period. However, for some unknown reason the country ceased to publish the overall systematic statistical data after 1965.[29]

In sum, the investment in heavy industry accounted for 63.7–83.3% during the 1954–64 period and for about 74.0% during the plan period (1961–70), while light industry occupied only 16.7%–36.3% during the 1954–64 period and about 26.0% during the plan period. The ratio of industrial sector to total social product value ranged from 23.2% in 1946, 35.6% in 1949, 30.7% in 1953, 40.1% in 1956, 57.1% in 1960, and 62.3% in 1963, to 57.3% in 1970, while the agricultural sector accounted for 59.1%, 40.6%, 41.6%, 26.6%, 23.6%, 19.3%, and 21.5% in the respective years. The remainder of the social product value came from the transport and communication, construction, commerce, and other sectors. The annual average growth rate for the industrial sector as a whole was set at 18%, but actual achievement was far short of the plan goal: 14% in 1961, 17% in 1962, 8% in 1963, 17% in 1964, and 14% in 1965.

During the plan period of 1961–70, the producer goods sector grew at an annual rate of 13.9% while consumer goods production increased at only 10.8% per annum. The poor actual achievement below the plan target was perhaps due partly to a misallocation of resources to the military sector, and partly to foreign aid cuts from the Soviet Union and China. Poor outcome appears to be at least part of the reason why North Korean officials decided to extend the plan period by three years.

(g) The Six-Year Plan: 1971–1976 The Six-Year Plan attempted to revise its self-reliance policy in favour of facilitating expanded access to foreign capital and technology in an attempt to overcome its problems. This plan proposed the so-called 'three technology revolutions'—(1) eliminating work-load gaps between heavy and light industries; (2) reducing gaps between farm incomes and industrial incomes; and (3) alleviating women's household burdens —all through the introduction and development of new technologies and methods.

North Korea employed three slogans to speed up the early accomplishment of the plan. These were named after the slogans employed in military combat: (1) speedy attack; (2) shock attack; and (3) exterminatory attack. This illustrates the kind of emergency measures taken to pave the way for the development of situations such as the North–South talks, held in 1972 for the first time since the nation's division nearly twenty years before.

Partly because of its all-out efforts and partly thanks to capital and technology imports from abroad, the plan targets were able to be met one and a half years ahead of the set time with annual growth rates of 10.4% (average industrial output growth rate 16.3%). The targets were set for income growth of 180%, industrial output growth of 220%, and annual grain production of 7.0– 7.5 million tons. Statistics reveal that actual industrial output increased 250%, machinery industry recorded a growth rate of 19.1% per year, labour productivity a growth rate of 155% during the plan period, and grain production reached 8.0 million tons in 1976. However, a rise in foreign imports resulted in the accumulation of foreign debts and subsequent problems in debt repayment capability during this open-door policy period (see Table 2.6). As the only communist country to default on its debts, North Korea has earned an unenviable reputation which has virtually cut it off

TABLE 2.6 North Korea's outstanding foreign debts, 1971–1989 (US$ million)

	Communist countries	Western countries[a]	Total
1971	250	17	267
1972	150	204	354
1973	109	379	404
1974	120	400	520
1975	70	130	200
1980	124	222	346
1983	119	156	275
1984	117	113	230
1985	156	134	290
1986	183	223	406
1987	241	280	521
1988	247	273	520[b]
1989	404	274	678[c]

[a] Japan, West Germany, Italy, France, Sweden, Australia, Canada, Singapore, etc.
[b] Economist Intelligence Unit, *Country Report: China, North Korea*, no. 1 (1990), 5.
[c] Preliminary.

Sources: 1971–4: *Le Monde* (France); 1975–87: South Korean National Unification Board, *North Korean Data*, no. 74 (1987), 47.

since 1975 from access to advanced Western technology. As a result, North Korea has turned once more to the *chuche* policy, although it badly needs foreign technology and capital to modernize its economy.

(h) The second Seven-Year Plan: 1978–1984 Putting aside the year 1977 as a buffer period, the North Korean government launched its second Seven-Year Plan in 1978. Major targets were: (1) to promote self-reliance, modernization, and technological development; (2) to strengthen the economy; (3) to promote exports of manufactured goods; and (4) to modernize the transportation sector.

This second Seven-Year Plan set growth targets no higher, and in some cases even lower, than those proclaimed earlier. Fulfilment of the plan was claimed in early 1984, with an average annual rise in industrial output of 12.2%. However, annual official figures of actual achievement *v.* plan targets have been few and far

TABLE 2.7 Trends of North Korean plan targets *v.* actual performance, by major commodities, 1961–70, 1971–77, 1978–84

Major commodities	Seven-Year Plan (1961–70) Targets	Performance	Six-Year Plan (1971–77) Targets	Performance	Second Seven-Year Plan (1978–84) Targets	Performance
Electricity (billion kwh)	17.5	16.5	28.0–30.0	28.0	56.0–60.0	(2 times over 1970)
Coal ('000 tons)	25 000	27 500	50 000–53 000	50 000 (in 1975)	70 000–80 000	70 000
Steel ('000 tons)	2300	2200	3800–4000	4000	7400–8000	—
Pressed steel ('000 tons)	1700	1980	2800–3000	—	5600–6000	—
Nonferrous metals	(copper 2.5-fold, lead 1.8-fold, zinc 1.6-fold)	overfulfilment	(copper 1.7-fold, lead 1.8-fold, zinc 1.8-fold)	—	1 million tons	—
Machinery	(3.3-fold)	(2.2-fold)	(2.7-fold)	30 000 units	50 000 units	(2.3-fold)
Fertilizer ('000 tons)	1700	1500	(2.0-fold)	3000 (in 1975)	5000	5000
Chemical fibres ('000 tons)	50	(2.4-fold)	(1.9-fold)	50	(1.8-fold)	(1.8-fold)
Synthetic plastics ('000 tons)	(68-fold)	(40-fold)	(3-fold)	50	100 (2.0-fold)	(2.4-fold)
Cement ('000 tons)	4300 (1.9-fold)	4000 (1.8-fold)	7500–8000	8000 (in 1976)	12 000–13 000	2.2 times over 1970 capacity
Textiles (m)	5 (2.6-fold)	4 (2.1-fold)	5–6	6	8	8
Cotton fabrics	—	—	Sweaters, jackets (2.1-fold)	Sweaters, jackets (6-fold)	Cotton (1.7-fold)	—
Shoes ('000 pairs)	40 700 (1.8-fold)	capacity increase	70 000	(2.4-fold)	(100 000)	—

Processed meat products	(4.8-fold)	(2.2-fold)	(5.9-fold)	—	—	—
Processed fruit products	(12-fold)	(15-fold)	(9.2-fold)	(1.8-fold)	—	—
Paper ('000 tons)	250 (5.3-fold)	113 (2.4-fold)	(1.8-fold)	(1.8-fold)	—	3500
Marine products ('000 tons)	1200	—	1600–1800	1600 (in 1975)	3500	—
Fishing[a] ('000 tons)	—	—	1300	—	2700	—
Grains ('000 tons)	6600	—	7000–7500	8500 (in 1977)	10 000	9000
Rice[b]	300	—	350	—	—	—
Tractors in use	70 000 (15 hp)	(3.3-fold)	(4.2-fold)	(4.0-fold)	10 units per 100 cheongbo more production	—
Farms with cars	13 000	(6.4-fold)	(2.5-fold)	1 car per 100 cheongbo	—	—
Fertilizer usage	570 kg per cheongbo	(3.2-fold)	1 ton per cheongbo	1.3 ton per cheongbo	2 tons per cheongbo	—
Pesticides usage	(3-fold)	(3.3-fold)	—	(2.2-fold)	—	—
Nuts ('000 tons)	(2.3-fold)	(2.0-fold)	800–1000	(2.2-fold)	1500	—
Eggs (millions)	800	700	3000	—	—	—
Meat ('000 tons)	350 (3.9-fold)	—	400	550	800–900	—
Silk ('000 tons)	22 (2.8-fold)	—	40	—	—	—

[a] Includes shellfish.
[b] Unpolished.

Source: Tama Siromuto, 'The Achievements of North Korean Economic Plan and Problems', paper presented at a conference in Seoul on 27 August 1989. The data in the table were obtained from North Korean government reports issued at various times.

between: an isolated absolute magnitude here and there, but nearly always a percentage increase (always an increase and no decrease) over a base-line which has not been assigned any quantity or value statistics. The case is particularly frustrating for the second Seven-Year Plan, since the North was so secretive about its economic and social statistics, as indicated in Table 2.7. Perhaps the failure to achieve its plan targets was due to the lack of technological modernization as well as to shortages of capital inputs.

Technological modernization over the decade 1968–77 had been made no easier by North Korea's inability to actually pay for anything it bought overseas. Therefore it is hard to refrain from concluding that the country's economic plan was stuck. In addition, North Korea had by now become a victim of its erstwhile success: the methods that had propelled it hitherto were now increasingly obstacles for the future, in the sense that marginal increases in productivity were more difficult to achieve, with workers being forced to work even harder with the same old machinery. Table 2.7 compares the plan targets with actual achievements by major commodity items for each relevant plan.

A notable occurrence during the second Seven-Year Plan was the announcement of 'ten major targets of communist economic construction' (hereafter 'ten major targets') at the Sixth Congress of the Korean Workers' Party held on 10 October 1980. These targets, which were to be achieved by the end of the 1980s, included 15 million tons of steel and 15 million tons of grain, and equally high targets for electric power, coal, non-ferrous metals, cement, chemical fertilizers, textiles, fishery products, and tideland reclamation (see Table 2.8).

Performance of the second Seven-Year Plan was comparatively good for the first three years, from 1978 to 1980, but industrial production has been extremely unstable since the Sixth Party Congress in 1980. The interlude of the 1980s' 'ten major targets' under such slogans as 'realization of communism in rice, textiles, etc.' might have slackened the progress of the second Seven-Year Plan. No interim figures have been reported for industrial production growth since 1981, but the second Seven-Year Plan targets and the 'ten major targets' together provide some grounds for estimating the annualized compound growth rates of the 'ten major targets' (see final column of Table 2.8).

In 1984, North Korea simply announced its completion of the

TABLE 2.8 North Korea's major commodities growth targets for 1980 v. the second Seven-Year Plan (SSYP) targets

	SSYP targets (announced at Supreme People's Meeting, 17 Dec. 1977) (A)	Ten major targets of Socialist Economic Construction (Sixth Congress of Workers' Party, 10 Oct. 1980) (B)	$\frac{(B)-(A)}{(A)} \times 100$ (C)	Annualized compound growth rates of (C) (6-year compound rates)[b]
1 Steel (million tons)	7.400–8	15	102.7–87.5	13~11
2 Nonferrous metals (million tons)	1	1.5	50	7
3 Coal (million tons)	70~80	120	71.4–50.0	9.5–7
4 Electricity (billion kWh)	56–60	100	78.6–66.6	10.5–9
5 Cement (million tons)	12~13	20	66.6–53.8	9–7
6 Textiles[a] (million metres)	800	1500	87.5	11
7 Marine products (million tons)	3.5	5	42.9	6
8 Chemical fertilizer (million tons)	5	7	40.0	6
9 Tideland reclamation (thousand ha)	100	300	200.0	20
10 Grains (million tons)	10	15	50	7

[a] In SSYP, this is classified as 'machinery-processed goods', the meaning of which is unclear, while in the 1980 targets it is classified as 'textiles'.
[b] 1978–84 average.

Source: as Table 2.7.

second Seven-Year Plan without giving any details of the achievements, and entered into a two-year adjustment period before
beginning the third Seven-Year Plan in 1987.

(i) The third Seven-Year Plan: 1987–1993   Originally, the new
Seven-Year Plan was to begin in 1985. However, two years of
adjustment, 1985 and 1986, were required before the new plan
could get under way. President Kim Il-Sung made the first official
announcement that the plan would be implemented in 1987 in a
speech on 20 October 1986 at a Pyongyang citizens' rally welcoming
Chairman Erich Honecker of the German Democratic Republic.[30]

It is not clear why they needed a two-year adjustment period
before formulating the new plan—perhaps because of the generally poor economic situation due to the disappointing achievements
of the second Seven-Year Plan. Unconfirmed rumours said that
there was a great deal of internal argument over where to lay the
blame and what direction to take in the new plan.

It is also worth noting that North Korea concluded a long-term
merchandise distribution and payment agreement with the Soviet
Union for the years 1986–90 in February 1986 and a long-term
agreement with the People's Republic of China for the years 1987–
91 in September of the same year. The prolonged negotiations for
these trade agreements might have been another cause of the delay
in preparing the plan.[31]

The basic task of the third Seven-Year Plan is to continue the
vigorous promotion of the '*chuche* idea', i.e. modernization and
scientization of the economy, and to lay a solid material and technical foundation for the 'complete victory of communism'. Except
for the phrase 'complete victory of communism', this is roughly
similar to the basic goal of the second Seven-Year Plan.

In the new plan, the need for a technical revolution is placed
at the top of the agenda. The emphasis on science and technology was demonstrated by President Kim Il-Sung in his New
Year message of 1986, entitled 'For the Complete Victory of
Communism':

Currently, the major goal of scientific and technical development in our
country is to achieve a complete technical restructuring of the people's
economy. Old, obsolete equipment in the people's economic sector must
be modernized, and production processes must be mechanized, automated,
computerized, and equipped with robots.

It is interesting to note that the leader of North Korea is well aware of the core of problems in the lagging economy: that is, the need to promote science and technology so as to induce a technical restructuring of the economy. While a technological revolution seems to be the key to future DPRK economic expansion, the persistence of the Northern leader in sticking to the *chuche* policy is surely an irony, because the development of science and technology in an isolated economy is likely to be hard to achieve.[32]

The overall targets for expansion during the period of the plan are shown in Tables 2.9 and 2.10. They are: (1) to increase industrial output 190% (an annual average of 10% growth), agricultural output 140%, social output 180%, and national income 170%; (2) to increase basic construction investment 160% over that of the second Seven-Year Plan; (3) to increase domestic retail goods distribution 210% and foreign trade 320%; and (4) to increase labour productivity in industry 140% and in construction 150%.

Since no figures were reported for agricultural output, social output, basic construction, or foreign trade under the second Seven-Year Plan, it is impossible to figure out the targets in terms of absolute quantities or values. But it seems that targets for the third Seven-Year Plan were set comparatively low. Under the second Seven-Year Plan, the target for industrial output was set for 220% growth (an annual average growth rate of 12.2%) and the national income target was a 190% increase.

Many outside watchers of the North Korean economy have had reservations about the possibility of accomplishing the plan targets, but surprisingly, on 30 May 1990 North Korea announced through its official news agency that the Korean working people in different fields of the national economy had fulfilled their assignments of the third Seven-Year Plan (1987–93). It reports that 'the working people across the country are waging an energetic drive for increased production and economies to carry out this year's plans before 10 October, the 45th anniversary of the founding of the Workers' Party of Korea, and the third Seven-Year Plan more than one and a half years ahead of the set time. . . . An increasing number of units and working people are beating the yearly and Seven-Year Plan targets ahead of schedule.'[33] Although the third plan sets targets lower than the second, early fulfilment of the plan is doubtful if not in quantity, then in quality. Plan targets may be fulfilled ahead of the set time, if this is essential, at least in terms

TABLE 2.9 General indices for North Korea's third Seven-Year Plan[a]

	Third Seven-Year Plan	Second Seven-Year Plan	
	Goals	Goals	Performance
Gross value of industrial output	1.9	2.2	2.2
(Average annual growth: %)	(10)	(12.1)	(12.2)
Producer goods	1.9 (9.6)	2.2 (11.9)	2.2 (11.9)
Consumer goods	1.8 (8.8)	2.1 (11.9)	2.1 (11.9)
Gross value of farm output	1.4 (4.9)	—	—
Gross value of social output	1.8 (8.8)	—	—
National income	1.7 (7.9)	1.9 (11.2)	1.8 (11.2)
Actual income of non-farmers	1.6 (6.9)	—	1.6 (6.9)
Actual income of farmers	1.7 (7.9)	—	1.4 (4.9)
Volume of rail freight	1.6	1.7	1.8
Volume of road freight	2.6	4.0	2.2
Basic construction investment[b]	1.6	—	1.5
Value of retail merchandise distributed	2.1	1.9	1.9
Value of trade	3.2	—	—
Industrial labour productivity	1.6	1.7	—
Basic construction labour productivity	1.5	1.6	—
Industrial cost reduction[c] (%)	3.4	3.7	—
Basic construction cost reduction[c] (%)	4.6	5.1	—

[a] Figures in parentheses are percentages of average annual increase and all other figures are multiplication factors of increase over index year.
[b] Compared with previous planning year.
[c] Annual average.

Sources: Gun-Mo Yi, 'Report on Third Seven-Year Plan' to Eighth Period 2nd Session of DPRK Supreme People's Assembly; DPRK Supreme People's Assembly Resolution: 'On the Third Seven-Year (1987–1993) Plan for Economic Expansion of the Democratic People's Republic of Korea'.

of quantity, but usually at the sacrifice of quality. There is, however, no way to follow up or to check the proclaimed fulfilment of the third Seven-Year Plan targets ahead of schedule. Yet scepticism about the claims of early fulfilment of the plan targets seems in order.

One interesting feature of the third Seven-Year Plan is an attempt to narrow the income gap between non-farmers and farmers. As seen in Table 2.9, the income of peasants is targeted to grow by 170%, while incomes of labourers and office workers rise by 160% during the third Seven-Year Plan period. But simple arithmetic calculation of target income levels of both non-farmers and farmers shows a considerable gap between the two, assuming initial equality of incomes (say, 100 alike for both) at the starting-point of the second Seven-Year Plan period.

The existence of a gap, taken into account in only the two consecutive plans, is due to the relatively higher growth of non-farmer's income (1.6% annually) than that of farmers (1.4% annually) during the second Seven-Year Plan period:

Labourers and office workers: $100 \times 1.6 \times 1.6 = 256$
Peasants: $\qquad 100 \times 1.4 \times 1.7 = 238$

This arithmetic simply implies that at the end of the third Seven-Year Plan the differential of incomes between the two groups is still far from narrowing: the income of non-farmers (whose income grows at the annual rate of 1.6%) would still be higher by a margin of 7.6% than that of farmers, whose income would grow at annual average rates of 7.9% during the plan period. Nevertheless, it is notable that North Korean policy-makers have admitted that there were problems of an incomes gap between farmers and non-farmers and that they face the necessity of narrowing the gap in the third Seven-Year Plan.

North Korea is also likely to pursue a more open position with the outside world. The only barriers to its open-door policy, if any, would include its outstanding foreign debts, its rigidity in economic management, its political burden of having to cement Kim Chong-Il's power base, and the rapid and severe collapse of other communist powers elsewhere.

Economic management and systems

In North Korea, economic management is fully combined with party leadership and political guidance.

TABLE 2.10 Major targets of the third Seven-Year Plan[a]

	Third Seven-Year Plan	Second Seven-Year Plan	
	Targets	Targets	Results
+Electric power (billion kWh)	100	56–60	—
+Coal (million tons)	120	70–80	70
+Steel (million tons)	10	7.4–8.0	—
+Nonferrous metals (million tons)	1.7	1.0	—
Machinery	2.5-fold	(5 million tons)	2.3-fold
+Chemical fertilizer (million tons)	7.2	5.0 (1.6-fold)	5.0
Chemical fibres ('000 tons)	225	(1.8-fold)	(1.8-fold)
Resin, plastics ('000 tons)	500	(synthetic resin 2-fold)	(synthetic resin 2.4-fold)
Sodium carbonate	4.5-fold	3.4-fold	—
Caustic soda	2.1-fold	1.8-fold	—
Sulphuric acid	3-fold	1.9-fold	—
+Cement (million tons)	22	12–13	production capacity 12
+Textiles (million metres)	150	80	80
Regional industry	2.5-fold	2.4-fold	—
+Grains (million tons)	15	10	10
Rice (million tons)	7	—	—

+Tideland reclamation ('000 ha)	300	100	—
No. of tractors per 10 ha	10–12	10	—
Chemicals per 1 ha (tons)	2.5	2.0	—
Meat (million tons)	1.7	0.8–0.9	—
Eggs (billion)	7	—	—
Fruit (million tons)	2.0	1.5	—
+Marine products (million tons)	11	3.5	3.5
Fish (million tons)	3.0	—	—
Seaweed, shellfish breeding (million tons)	8.0	—	—
Residential construction ('000)	150–200 per year	200–300 per year	several hundred thousand during period
Technicians and specialists (millions)	2.0	—	1.25
Number of preventive medicine groups	1.2-fold	1.3-fold	290 new
Number of hospital beds	1.3-fold	1.2-fold	—
Doctors per 10 000 people	43	—	—

a + indicates 'Ten major targets (1981–90)'. Reference values not amenable to direct comparison are given in parentheses.

Source: North Korean Official Reports.

The great leader Comrade Kim Il-Sung put forward the unique idea that the collective guidance of the party committees should be made to play the role of masters in economic management during the plan period in order to carry on economic management in keeping with the intrinsic requirement of the communist system.[34]

In December 1961 Kim Il-Sung created the so-called 'Taean Work System' in addition to a new system of agricultural guidance, and thus established the *chuche*-based economic management system intended to carry on economic management by the masses under the collective guidance of the party committee.[35]

Since the vitality of the *chuche*-oriented economic management system lies in recognizing the worth of the economic and technological guidance of the Workers' Party Committee, the administration can be firmly relied upon to follow the party's leadership.

The Workers' Party has about 27 divisions (as of 1989) under its secretarial office, of which 11 are charged directly with economic management and operation. Occasionally, the secretarial meetings of the Workers' Party Assembly, the Party's Central Committee, or the Political Bureau of the Party's Central Committee either adopt new economic policies or recommend major goals or directions of economic plans. In accordance with North Korea's constitution, the Central People's Committee is the supreme leadership organ for the People's Democratic Republic of Korea. This Committee is in charge of providing guidance and leadership to the overall administration programme of the government. The President of DPRK holds the chairmanship of the Central People's Committee, which in turn consists of members of the political bureau of the Workers' Party.

The Office of State Affairs under the Central People's Committee consists of 15 commissions, 21 ministries, and 1 board (at the end of August 1990), of which non-economics-related ministries and/or commissions are the ministries of foreign affairs, education and culture, health and medical affairs, and social security affairs, and the State Sport Commission. As such, there are many overlapping commissions and ministries which are directly related to economic management, and their functionings are also diversified by different industrial sectors.

At the local level, two organizations—the party and the administration—deal with economic management. The party has hierarchical party committees at province, city, and county level, while the administration runs its people's committees at province, city, and county level, its administration and economic guidance committees and rural accounting committees at province level, its administration committees at both city and county level, and its co-operative management committees at county and village level. The local-level organizations, whether party or administration, are closely guided by the central organizations.

(a) Industry management   In North Korea, industrial plants and enterprises are classified as either state plants and enterprises or co-operative plants and enterprises, by ownership type, and as either central or local, according to location. They are also classified as single enterprise, union enterprise or complex enterprise, according to management type. Finally, the central plants and enterprises are classified as special, first-, second-, or third-class according to their size. Here the unit of industrial management, which is related to production activities under the state plan, is called a 'plant and enterprise'. A large plant or enterprise whose production lines are interrelated geographically is called a complex plant or enterprise.

For plants and enterprises of above third-class size, the Central Party and the Economic Bureau of the Office of State Affairs are responsible for both guidance and control, while local party committees and administration and Economic Guidance Committees at the provincial level are assigned to guide and control the local plants and enterprises.

The wage structures for plant and enterprise workers are based on independent pay schedules and tables in accordance with job types, number of years worked, and individual working abilities. With reference to the average basic wage scale, a bonus (reward) system is applied in cases of overfulfilment of the plan target, as explained in Section 2.1 above.

(b) Agricultural management   In the agricultural sector, economic management is conducted in a unit of each village (*Ri*) co-operative jointly by a County Farm Management Committee and a Provincial Farm Accounting Committee in co-operation with the Central State Agricultural Committee. The County Farm Management Committee was established in 1962 to provide farmers with professional guidance. It acts as a farm enterprise, whose function is to control and operate all co-operative farms within each county and all state enterprises in the agricultural sector. The basic unit of production in a co-operative farm is the working unit. Each working unit is allocated a land plot and farm tools so as to fulfil annual production assignments.

Annual settlement of accounts in each co-operative farm is made both with products in kind and by cash in proportion to each farmer's contribution to overall production. The distribution of cash income is made to participating farmers from sales to the government after deduction of both the co-operative farm tax and a donation to the co-operative accumulation fund; the produce that is left over after government purchase is allocated to farmers as payment in kind. Preferential allocation is made to any farmers overfulfilling their plan targets.

(c) Plan formulation procedure   The formulation and execution of economic plans in North Korea are truly a co-operative venture between the State Planning Commission, the Office of State Affairs, and local committees of each factory and enterprise. As in other communist countries, the State Planning Commission is entrusted with the formulation, execution, and control of economic plans and policies for the Workers' Party in North Korea.

The stages of plan formulation in North Korea are as follows.

First Stage: Formulation of the preliminary plan   Each factory, enterprise, and co-operative farm formulates its own plan and submits it to the local administration and planning commission, which takes all submitted plan forecasts for review and consolidation. Then the economic bureaux of the Office of State Affairs collect all forecasts to submit to the State Planning Commission.

Second Stage: Formulation of control   The State Planning Commission formulates sectoral targets and sets up control forecasts based on the preliminary plan forecasts submitted by all local factories and enterprises. These then go to the economic branches of the relevant administrations.

Third Stage: Formulation of draft plan   Based on the control estimates set up by the State Planning Commission, all administration branches, factories, and enterprises formulate their respective plans and re-submit them to the State Planning Commission. The State Planning Commission then finalizes overall draft plans.

Fourth Stage: Decision-making and approval   The Central Party Committee and Office of State Affairs meet together to overview the details of draft plans, which are then submitted to the Supreme People's Assembly for final approval.

(d) Public finance Public finance plays an important role in assuring balanced development and equilibrium in both wealth accumulation and distribution in the planned economy. The central government budget and local government budget are consolidated into the state (national) budget. The central government collects and dispenses its budget for society as a whole, while the local budget is collected from and used for local administrative organizations, local factories, enterprises, and welfare facilities. The central budget accounts for about 85% of the national (state) budget, while the local budget accounts for 15%, although the relative proportion of the local budget has been increasing in recent years.

The central government budget is financed mostly from the net income of the central government (which consists of transaction revenue,[36] social insurance fees, income taxes imposed on cooperatives, and profits of state-run enterprises); from transfers from local governments; and from other revenue sources. The expenditure items include expenses for economic development, military expenses, expenditure on social and cultural facilities, and administration expenses.

The local government budget revenue comes mostly from the revenue sources of local industries, welfare and service sector revenues, and subsidies from the central government. Its expenses consist of the cost of local business and administration agencies, welfare services, and transfers to central government. The plan and formulation of the annual budget rest with the Finance Department of the Office of State Affairs in co-operation with both the State Planning Commission and the Price Control Committee. The preparation of the budget begins in July of the preceding year and is completed in March of each budget year. This budget plan needs to be approved by the Supreme People's Assembly held annually in April, but the regular budget year runs from 1 January to 31 December. The execution of an operating budget between January and April every year is usually based on pre- or post-approval.

(e) Banking The important function of banking in North Korea is to control the planned uses of both physical and monetary resources in every sector of the economy. This is termed 'the control by *won* (money)'. All official agencies, factories, enterprises, shopping centres, etc., have to keep open accounts with designated banks so that every monetary transaction will occur through the intermediation of banks. However, ordinary people make banking transactions merely when depositing and withdrawing individual savings.

Major banks in North Korea include the Chosen Central Bank, the Foreign Trade Bank, the Kumkang Bank, and the Chosen Daesung Bank, all dealing with payments to foreign traders. The main functions of each bank are summarized as follows.

The Central Bank of the People's Republic of Korea This is the central government bank of North Korea. It is in charge of issuing notes and coins, ensuring monetary control, taking receipts and payments of government taxes, supplying operating funds to government agencies and businesses, and supervising the activities of the Foreign Trade Bank, the Kumkang Bank, and the Daesung Bank. It also conducts trade settlements with other socialist and non-allied member countries according to mutual agreements.

The Foreign Trade Bank The Foreign Trade Bank is responsible for extending settlement arrangements for international transactions and credits and also for transacting payment guarantees. In addition, it handles foreign exchange business and maintains contracts with banks abroad. Lastly, this bank is in charge of issuing foreign exchange notes to foreign visitors to North Korea.

The Kumkang Bank This bank was first established in September 1978 to meet increasing demands for an international trade settlement facility, particularly in relation to exports and imports of machinery, steel and metal products, iron ore, and chemical products.

The Chosen Daesung Bank Like the Kumkang Bank, this bank was set up in November 1978 to assist the Chosen Daesung Trade Company, the Chosen Donghae Marine Transportation Company, and the Chosen Mankyoung Trade Company in their export and import businesses and their settlement of trade accounts, particularly concerning precious metals and foreign exchange. This bank has its correspondent bank, the Golden Star Bank, in Vienna.

(f) Marketing The marketing of products in North Korea occurs in accordance with the state plan. The types of marketing are classified according to different ownership types, for which differing price systems apply. For example, in the case of a transaction between two state-run enterprises, the relevant wholesale price applies, while for trade between a state-run enterprise and a co-operative organization, the so-called industrial wholesale price applies. Different price mechanisms are used in accordance with different transaction types, reflecting in turn different ownership types such as state ownership, co-operative ownership, and private ownership. The ownership relation can be changed as the consequence of trade, of course. The types of market are divided into state-run markets, co-operative markets, and farmers' markets according to the relevant market-place in which trade predominantly takes place. In the state-run market, wholesale trade of industrial products dominates and occasionally retail transactions also occur. The co-operative market includes the farmers' co-operative market, the producers' co-operative market, and the marine products co-operative market, not to mention barbershops and public bath-houses. At a farmers' market, farm products raised in private plots and homemade by-products are purchased and sold.

(g) Prices The price system in North Korea functions to represent the labour inputs (socially needed amounts of labour time) embodied in commodity outputs; to stimulate the production expansion; and to adjust the demand and supply of goods. Of these three functions, the adjustment of demand and supply—that is, the price-planning function—is most important. In North Korea, a commodity's price is determined not by the demand and supply mechanism, but arbitrarily by the central planners so as to control the demand for and supply of commodities. Therefore, the price of a commodity is not relevant in many cases to the actual value of that commodity. Nevertheless, any commodity price is regarded as theoretically representing the average amount of labour inputs into that commodity. Based on so-called socially needed labour time, which represents the average quantity of labour put into the production of the good, the price is determined so as to raise labour productivity, increase cost effectiveness, and reflect savings in the use of raw materials and consumers' living standards.

Basically, there exist three categories of prices in North Korea. The first is the wholesale price of goods leaving factory or enterprise, which in turn is subdivided into the firm wholesale price and the industrial wholesale price. The firm wholesale price is a price based on both the producer's cost (direct cost plus depreciation cost plus wage and administration cost) and net income (profits) of the firm. This price applies to the mutual trade of the producer's goods between the same types of factory and enterprise. The industrial wholesale price is formed by adding the transaction revenue (which is similar in nature to a sales tax) to the firm wholesale price. This price applies to the trade between different types of ownership factories and enterprises.

The second category is the retail price of commercial traders. This price consists of marketing costs (transportation, wages and salaries, and packaging costs) and profits in addition to the industrial wholesale price. It applies to any consumer good purchased at a retail shop.

Thirdly, the farmers' market price is a flexible price to be determined in any farm products market in accordance with the demand and supply conditions prevailing at the time.

(h) Money in circulation   In most communist countries money is generally regarded as a special commodity which functions to value each commodity in question; such functions as means of exchange, value stores, and savings are also widely acknowledged. But in North Korea the function of money is limited to serving only as a means of exchange. The practice of arbitrary manipulation of a commodity price obliterates the real value of the commodity. The central planners usually control the demand side of the economy, not to mention manipulating the circulation of cash to control business activities. All these practices muddle the generally defined functioning of money in the economy. North Korea calls it 'the control by *won*', and maintains a state of material equilibrium by adjusting the amount of money in circulation in the economy.

There are two sorts of monetary measure in North Korea. The first is the amount of cash in circulation to make the market clear for daily transactions such as commodity purchases, wage payments, etc. The other one is just the accounting or book money that is used to clear transactions among factories and enterprises, the state provision of loans to businesses, businesses' payments to the state, etc. This occurs in mere transfers of money between bank accounts of the parties involved.

The species of commodity money in North Korea consists of coins (denominations: 1, 5, 10, and 50 *chon* (*jun*), where 100 *chon* (*jun*) is equal to one *won*, and paper notes (denominations: 1, 5, 10, 50 and 100 *won*). There are also foreign exchange notes equivalent to coins and notes in the above denominations. These are for the exclusive use of foreigners visiting North Korea.

(i) Interest rates   Pre-fixed interest rates on both bank loans and savings exist in North Korea. As in the case of commodity prices, the interest rates are determined by the plan authority without regard to the supply and demand conditions of the loan market. The rates are usually kept very low.

(j) Foreign exchange rates   Technically, North Korean currency is transacted with foreign currencies at officially determined exchange rates. The exchange rates are arbitrated in accordance with the 'foreign trade plan' or 'foreign exchange plan'. Three different multi-exchange-rate systems exist currently in North Korea.

1. *Official rate or basic rate*   How the official or basic rate is determined is not known, but this rate seems to be determined by taking into account exchange rates between the Soviet currency and Western currencies and is linked to Soviet roubles at a parity of Rb1 = W1.34. The official rate, which values its domestic currency relatively highly, is used to convert North Korea's income into foreign dollar income for release abroad.

2. *Trade exchange rate*   The trade exchange rate (commercial rate) is applied to international transactions with the external world. This rate in theory represents the ratio of domestic prices to foreign prices, but where price systems differ between the countries being compared it fails to represent correctly the relative purchasing powers of the currencies involved.

The Foreign Trade Bank revises the trade exchange rates from time to time.

3. *Non-commercial rate*   Theoretically, this non-commercial rate of exchange is determined on the basis of the household expenditure ratios of any two countries. However, in North Korea this rate simply represents the trade rate plus a certain premium arbitrarily set by the government and has been $1 = W2.128 since February 1973. This rate applies to invisible foreign trade and capital transactions. Recently, however, it has come to be replaced by the trade (commercial) exchange rate. Table 2.11 shows the North Korean *won*–US dollar exchange rates.

TABLE 2.11 Exchange rates in North Korea, 1960–1991

	Official rate[a]	Trade rate (W/$)[b]	
1949–59		2.20	
1960	1.20	2.20	
1961	1.20	2.20	
1962	1.20	2.20	
1963	1.20	2.20	
1964	1.20	2.20	
1965	1.20	2.20	
1966	1.20	2.20	
1967	1.20	2.57	
1968	1.20	2.57	
1969	1.20	2.57	
1970	1.20	2.57	
1971	1.11	2.36	(2.36)
1972	1.11	2.36	(2.36)
1973	1.11	2.36	(2.06)
1974	0.96	2.36	(2.15)
1975	0.96	2.05	(2.02)
1976	0.96	2.15	(2.02)
1977	0.96	2.15	(2.02)
1978	0.93	1.86	(1.86)
1979	0.84	1.79	(1.79)
1980	0.86	1.70	(1.77)
1981	0.92	1.77	(1.85)
1982	0.97	2.12	(2.00)
1983	1.02	2.18	(2.13)
1984	1.20	2.36	
1985	1.07	2.43	
1986	1.02	2.23	
1987	0.94[c]	2.14	
1988	0.94[c]	2.10	(2.26)[d]
1989	0.97[c]	2.10[e]	
1990	0.96[c]	2.08[e]	
1991	1.05[e]	2.15[e]	

[a] Official rates during 1949–59 were known to be in the range of W0.95–W1.20 per US dollar, but are not available on year-to-year basis.
[b] Numbers in parentheses are US CIA estimates as end-of-year rate: non-commercial rate to 1977, but 1978 is a special rate for March 1978 as reported. See CIA, *Handbook of Economic Statistics 1984*, 32; also Economic Intelligence Unit (EIU), *Country Report: China and North Korea*, each issue.
[c] Rate at March of the respective year, estimated by EIU.
[d] Tourist (non-commercial) rate.
[e] Estimates.

Sources: Official Reports; DPRK Trade Bank; US CIA estimates; and Economic Intelligence Unit estimates.

2.2.2 South Korea

When that day comes
Mt Samgak will rise and dance,
the waters of the Han will rise up.

If that day comes before I perish,
I will soar like a crow at night
and pound the Chongno bell with my head.
The bones of my skull
will scatter, but I shall die in joy.

When that day comes at last
I'll roll and leap and shout on the boulevard
and if joy is still stifled within my breast
I'll take a knife and skin my body
and make a magical drum and march with it
in the vanguard. O procession!
Let me once hear that thundering shout,
my eyes can close then.

'When that Day Comes' by
Shim Hun (1904–37)[37]

South Korea in transition

(a) From chaos to civil war: 1945–1953 A unifying dream, if any, held by Korean people in the poverty-stricken days after its independence and division was nothing more than to build their nation up from the ashes. But the path to reconstruction was thorny. The southern half of the divided nation was left with no workable resources and capital except for its unemployed illiterate population. Even before the euphoria at independence had faded away, people found themselves in turmoil, followed by civil war. From 1945 to 1948 the division of the country, the existence of a million or more jobless people, continuing conflicts between leftist and rightist groups, and the fumbling management by temporary US military administrators inhibited any economic growth. The then infant government of Syngman Rhee, harassed by guerrillas from the North on the one hand and by internal power struggles among leaders of different factions on the other, had little opportunity to build or foster the economy before it came under all-out attack from the North.

The war started on 25 June 1950. The three-year conflict caused more than a million deaths in South Korea and the division of many families. It not only destroyed much of the country's industry, housing, and other infrastructure, but also planted deep antagonism and distrust among those who had fought in the war. To make matters worse, there was a severe famine in 1951 and 1952, bringing people to the edge of starvation. In South Korea, the American Military Government estimated earlier that the average daily calorie intake fell from a little over 2000 in 1932–6 for all Korea to a little under 1500 in 1946–7 because of a shortage of food supplies. The level may have dropped still further, to less than 1000 in 1950–3. I myself, who was 11 years old when the war broke out, have vivid memories of being very hungry and of many people starving to death, adding to the war casualties. Those who survived the famine years filled their stomachs either with edible grasses and foots that pigs and domestic animals ate, or with leftovers from the foreign military compounds, or, if lucky enough, with food aid from the United States.[38] The survivors cannot look back on those days of agony without tears. How could they forgive those who started the war? But the feeling is changing now with the emergence of the post-war generation.

The war also brought a sharp drop in industrial production in 1950–2. By 1953 industrial output began to increase slightly as the war gradually drew to a stalemate.[39] Although accurate statistics on economic trends between 1945 and 1953 are lacking, economic output in the South in 1953 was probably not much more than one-third of the level of 1940.[40]

With regard to external trade, South Korea's exports and imports in 1946 were negligible. By 1949 they were still quite small—about $17 million and $22 million, respectively.[41] However, in spite of the war, exports in 1953 exceeded the 1949 level by more than 32% while imports were almost six times greater than in 1949. Nearly all South Korea's exports during this period were primary products including tungsten, graphite, copper, kaolin, and talc. Major imports were food grains and manufactured goods.[42]

Although South Korea adopted a market economy, the government played a large role in nearly all sectors of the economy from the very beginning. The foreign trade section was no exception. Private foreign trade was almost non-existent apart from a small amount of private barter during the 1945–53 period.[43] The government was a major exporter and importer both before and during the war. Various schemes for private export promotion were used during this period: a deposit system to avoid exchange risk; an export–import link system; direct export subsidies; a variety of preferential loans for exporters and export producers; and tariff exemptions. Nevertheless, the balance of trade deficits became greater during this period. Nearly all imports were financed by grant assistance or *won* redemptions by the UN Command during the civil war. Foreign relief and aid imports during the war were substantial, as can be seen in Tables 2.12 and 2.13. The total amount of relief and aid imports gradually increased from $58 million in 1950 to $194 million in 1953. The sources for relief and aid imports during the 1950–3 period were the US Economic Co-operation Administration (ECA) and its successor (in 1953), the International Co-operation Administration (ICA); Supplies from Economic Co-operation (SEC), administered through the UN Civil Assistance Command in Korea (UNCACK); the Civil Relief in Korea (CRIK); and the UN Korea Reconstruction Agency (UNKRA). The major items of relief and aid imports were food, clothing, medical supplies, fuel, industrial raw materials, and machine parts.

(b) Land reform in South Korea[44]   In contrast to farmland reform in North Korea, carried out in 1946, South Korea initiated its land tenure reform in early 1950. At the time of liberation in 1945, the land tenure relationship in South Korea was one of high tenancy and large numbers of landlords, reminiscent of a semi-feudalistic system. There were 2 065 477 farms, of which 300 000 (about 14%) were owner-operated while 720 000 (35%) were part owner-operated. The number of tenant farms was more than 1 million (49%) of the total farms. By area, approximately 850 000 hectares, or 36.6% of the total arable area of 2 053 811 hectares, was owner-operated;[45] the remainder was farmed by tenants. In the period immediately following liberation, the tenant farmers paid high rents, largely to absentee landlords who made up a small but wealthy aristocracy. There was an increasing discontent on the part of these farmers. Some landowners saw the inevitability of future land reform and voluntarily sold their farmland between 1945 and 1949. However, the abuses of the traditional tenant system continued, and voices were raised among tenant farmers and a group of economists for a comprehensive and profound land reform. In order to deal with the political instability which had reached a climax in rural areas, to improve the structure of rural

TABLE 2.12 South Korea: major statistics, 1945–1953

	Av. production index of major commodities[a] (1946 = 100)	Exports[b] (W million)	Imports[c] (W million)	Seoul wholesale price index (1947 = 100)	Foreign aid received[c] (US$ '000)
1940	—	0.48	0.77	0.4	—
1945	—	—	—	—	4 934
1946	100	0.05	0.16	55.0	49 496
1947	155	1.11	2.09	100.0	175 371
1948	184	7.20	8.86	162.9	179 593
1949	259	11.27	14.74	222.8	116 509
1950	180	32.57	5.21	348.0	58 706
1951	135	45.91	121.83	2194.1	106 542
1952	270	194.96	704.42	4570.8	161 327
1953	392	398.72	2237.01	5951.0	194 170

[a] Unweighted indices of commodities including rice, wheat, and barley, anthracite coal, tungsten ore, salt, processed marine products, cigarettes and tobacco, raw silk, cotton cloth, paper and paper products, laundry soap, cement, chinaware, nails, transformers, light bulbs, and electric power. For growth of each of these commodities during the period, see table 2.2 in Charles R. Frank, Jr, Kwang-Suk Kim, and Larry E. Westphal, *Foreign Trade Regimes and Economic Development: South Korea* (New York: Columbia University Press, 1975), 9.

[b] Exports and imports in current prices valued in *won* according to f.o.b. export or c.i.f. import prices until March 1951; thereafter, according to domestic market price. (Tariffs, domestic taxes, and trade margins were excluded from domestic prices to estimate the price of imports).

[c] Includes foreign relief goods and economic aid received from US GARIOA (Government Appropriation for Relief in Occupied Areas), ECA (Economic Co-operation Administration), ICA (International Co-operation Administration), CRIK (Civil Relief in Korea), and UNKRA (UN Korea Reconstruction Agency).

Source: Bank of Korea, *Economic Statistics Year Book* (1956, 1961).

TABLE 2.13 Foreign relief goods and economic aid received by South Korea, 1945–1953 (US$ '000)

| | | US Government | | | | |
	Total	GARIOA[a]	ECA and SEC[b]	ICA[c]	CRIK[d]	UNKRA[e]
1945	4 934	4 934	—	—	—	—
1946	49 496	49 496	—	—	—	—
1947	175 371	175 371	—	—	—	—
1948	179 593	179 593	—	—	—	—
1949	116 509	92 703	23 806	—	—	—
1950	58 706	—	49 330	—	9 376	—
1951	106 542	—	31 972	—	74 448	122
1952	161 327	—	3 824	—	155 534	1 969
1953	194 170	—	232	5571	158 787	29 580

[a] Government Appropriation for Relief in Occupied Areas.
[b] Economic Co-operation Administration and Supplies from Economic Co-operation. The ECA and SEC were absorbed into ICA in 1953, and then ICA was transformed into US AID in 1954.
[c] International Co-operation Administration, which was consolidated into US AID (Aid for International Development) in 1954.
[d] Civil Relief in Korea.
[e] UN Korea Reconstruction Agency.

Source: Bank of Korea, *Economic Statistics Yearbook* (1956), 249.

society by mitigating class conflicts, and to promote agricultural productivity, it was widely hoped to undertake farmland reform, including the abolition of tenant farming. Therefore, in March 1950 a final bill for farmland reform was passed by the National Assembly and promulgated as Law no. 108.

Article I of the 1950 Land Reform Law stated its intent as follows: 'on the basis of the Constitution of the Republic of Korea pertaining to farm lands: to improve the living conditions of farmers, to keep the balance of, and to develop, the national economy by increasing agricultural productivity'.

An important provision of the Law was that farmland owned by individuals other than farmers, farmland not owner-cultivated, and farmland exceeding the upper ceiling of 3 cheongbo (7.35 acres, or 2.974 hectares) were subject to government purchase and redistribution. By the time the reform was ready to be implemented, the

TABLE 2.14 Tenure patterns in South Korea, 1945–1965 (%)

Type of tenure	1945	1947	1965
Full-owner	13.8	16.5	69.5
Part-owner	34.6	38.3	23.5
Tenant	48.9	42.1	7.0
Slash-and-burn farmers	2.7	3.1	
Total farm households	2 065 000	2 172 000	2 507 000

Source: Bank of Korea and Ministry of Agriculture and Forestry, *Special Farmland Survey* (1966).

Korean War had broken out; consequently implementation had to be postponed, except in the Kyongsang Namdo provincial area where the government reform continued.

Despite the loss of documents related to the programme in the early phase of the conflict, the farmland reform was reinitiated after Seoul was recovered in September 1950. On 10 November 1950, land committees were reorganized to investigate the status of land for which official records had been lost. The reasons for continuing with the land reform, even during the conflict, were (1) to utilize land securities in selling vested enterprises, thus generating funds for war financing; (2) to demonstrate the emancipation of tenants from landlords; and (3) to use the rice collected for military provisions.

The land reform work was continued until 1969, and extended to cover the area north of the 38th Parallel that had been recovered from North Korean rule. However, the reform could not completely eradicate either the excess ownership problem or the tenancy problem, because some landowners succeeded in concealing their ownership of farmland, partially or completely, either by illegally evading the 3-cheongbo limit or by dividing land among relatives. While 31% of the land tentatively identified for land tenure reform was never officially designated or sold to tenants or other farmers, and while tenancy was never completely eradicated, the land reform in South Korea was considered to have achieved its major objectives of reducing the then political and social unrest and eventually increasing agriculture and labour productivity. Table 2.14 shows the trend of tenure patterns (1945–65), and Table 2.15

TABLE 2.15 Crop yields per cheongbo in South Korea (rough grain basis), 1937–8 to 1965–9[a]

Kind of crop	1937–8[b]	1945–9	1950–4	1955–9	1960–4	1965–9
				('000 kg)		
Rice (rough)	3.0	2.5	2.6	3.7	4.1	4.2
Common barley	1.5	0.9	1.0	2.0	2.2	2.3
Naked barley	1.6	1.0	1.1	1.6	1.7	2.5
Wheat	1.1	0.8	1.0	2.4	2.8	2.9
Italian millet	1.4	0.5	0.8	0.6	0.7	0.9
Total fertilizer supplies						
Nutrient basis (billion tons)	106	82	99	218	263	445

[a] 1 cheongbo = 2.45 acres. (To convert acres into hectares, multiply by 0.4046856.)
[b] Includes North and South Korea.

Sources: Ministry of Agriculture and Forestry (MAF), *Yearbook* (1969); *MAF Fertilizer Handbook* (1969).

TABLE 2.16 Foreign assistance received by South Korea, 1954–1961

	Total ($ '000)	US AID[a]	CRIK	PL 480 Title I	UNKRA
1954	153 925	82 437	50 191	—	21 297
1955	236 707	205 815	8 711	—	22 181
1956	326 705	271 049	331	32 955	23 370
1957	382 893	323 268	—	45 522	14 103
1958	321 272	325 629	—	47 896	7 747
1959	222 204	208 297	—	11 436	2 471
1960	245 394	225 237	—	19 913	244
1961	199 245	154 319	—	44 926	—

[a] US AID assistance consists of project assistance, non-project assistance, and technical co-operation. ICA aid was changed to AID aid.

Sources: Economic Planning Board, *Statistical Yearbook* (1962), 264; Bank of Korea, *Statistics Yearbook* (1962).

shows the trend of yield per cheongbo by crop (rough grain basis), illustrating the degree of land reform and the effect of land tenure reform on production and productivity, respectively.

(c) Post-war reconstruction: 1953–1961   The Armistice took effect on 27 July 1953. War damage to industrial offices, plants and equipment, public facilities, private dwellings, and transport equipment (exclusive of military installations) in South Korea was estimated to be about $3.0 billion. This was almost equal to the estimated GNP for 1952 and 1953 combined.[46]

Following the Korean War, economic growth was a secondary objective of the Syngman Rhee government from 1953 to 1960. During the first four years the government concentrated on the reconstruction of the country's infrastructure and industrial facilities that had been destroyed by the war. Thereafter, its policy emphasis shifted from the repair of war damage to the stabilization of prices. And the maintenance of strong military forces to deter any new attack from the North has been pursued consistently ever since then, absorbing about 4%–5% of GNP.

Both the reconstruction and the stabilization programmes during the post-Korean-War period were largely financed and aided by US and UN assistance (see Table 2.16). Beginning in 1953, a large amount of aid-financed imports arrived for relief and reconstruction. During 1953–60, total assistance provided by the UN Reconstruction Agency (UNKRA) amounted to approximately $120 million, and official US assistance reached approximately $1745 million including $158 million of PL 480 goods.[47] Foreign assistance was important for the reconstruction programmes in South Korea not only because it provided necessary imports, but also because the 'counterpart funds' generated by the sales of grant-aid dollars provided a non-inflationary source of domestic currency financing for investment. During 1954–60, the counterpart funds generated by sales of agricultural products supplied under Public Law 480 accounted for one-half of the government's budget.[48]

Despite its heavy dependence on US aid, Syngman Rhee's government had sharp disagreements with the US government over economic policies. During the entire post-war period, all major economic policy decisions, even those not directly involving aid funds, had to be made jointly by the South Korean government and the US aid mission to Korea (Office of Economic Coordinator). The areas that the government wished to give priority for economic policy often conflicted with those of the US aid mission. For example, the South Korean government wanted new factories and heavy industry financed by foreign aid, whereas the US government favoured light industry and better use of existing facilities. The United States urged the South Korean government to make the country more self-reliant, while the Koreans pressed for more aid. They differed also in the devaluation policy of the currency and in other areas such as tariffs and agricultural price policies.

Despite a large inflow of US economic aid and the provision of economic advice by US specialists, the overall rate of economic growth of South Korea, which was suffering a relatively high inflation rate during this period, was discouraging. As comparisons reveal in Chapter 3, economic growth in South Korea lagged behind that of North Korea during this period.[49]

The staggering economy and dictatorial government of Syngman Rhee became an increasing source of social and political dissatisfaction and unrest towards the end of the 1950s.[50]

During 1953–61 the South Korean economy was largely a traditional agricultural economy, with about two-thirds of the working population still engaged in the primary sector. The industrial policy pursued during this period may be loosely characterized as that of import substitution of non-durable consumer and intermediate goods under high tariffs and quota restrictions. However, the limited size of the domestic market and the shortage of capital supply became bottlenecks in the development strategy based on import substitution. The annual average growth rate of GNP during the periods 1953–61 and 1954–61 was only 3.7% and 4.0% respectively, and that of per capita GNP was a meagre 0.7%. Commodity exports remained negligible throughout the period, usually amounting to less than 1% of GNP, because a persistently overvalued domestic currency thwarted the export potential of the economy.

The South Korean economy in the post-civil-war period was characterized by the vicious circle of poverty; that is, low incomes led to low savings, which led to low investment, which led to low production, and back to low income (see Table 2.17). It suffered seriously from political instability, rampant inflation, and an inability to meet most of the basic needs of consumers. The rapid growth of population, accelerated urbanization, and unemployment further complicated South Korea's early economic problems. Population growth reached a rate of about 3% a year after the baby boom period which began in 1954. In addition, the migration of millions of people from the North during the war added to already existing socio-economic problems, including low capital formation, the shortages of both consumption goods and housing units, increased unemployment, and socio-political unrest.

In spite of these problems, however, changes were taking place in the economy in favour of tremendous growth potential. Thousands of men completed their military service and returned to civilian society experienced in functioning in a highly disciplined organization. South Korea's greatest resource is its people. It was its highly motivated and educated workers that were to enable the country to improve when a growth-oriented government came to power. South Korea's accomplishments in developing human resources early, despite low per capita incomes, made it possible to lay a solid foundation for subsequent economic development. The South Koreans have demonstrated an awareness of the importance of education as well as a great commitment to educate their next generation even at a substantial financial cost. Without the belt-tightening sacrifice of the then poor Korean parents for their children's education, the foundations for subsequent high growth in the 1960s and 1970s would never have been laid.

Table 2.17 illustrates the key statistics of the South Korean economy over 1954–61.

Economic Plans: Goals and Performances

> I took it as my main task, as the leader of the nation during this period, to inspire the courage and confidence of the people to achieve these national goals in a spirit of unity. In assuming this task, I found it necessary to reflect deeply on the history of our nation, on the characteristics of our culture and traditions, and on the capabilities of our people. I found in my reflections . . . what I feel are the common threads and the unique national spirit that link our past, our present and our future. Our past history seems at first glance to be more a record of misfortune than glory, but we also find in our past a strong inspiration, and we value even the misfortunes for the strong sense of determination they have nourished in our people's hearts.
>
> *Park Chung Hee's policy speech of his own commitment to the First Five-Year Plan*[51]

The first efforts towards economic planning in South Korea were made during the Korean War by the foreign aid agencies that were trying to assess the potential costs of rehabilitating the country's economy. Robert R. Nathan and Associates prepared an overall development programme in March 1954 for the UN Korean Reconstruction Agency, but this programme was not formally adopted by the Syngman Rhee government, mainly for political reasons.

A second planning effort was initiated in 1958 by the newly established Economic Development Council (which was transformed into the Economic Planning Board in 1961) of the Korean government. This was to be a seven-year plan which would be divided into a three-year and a four-year phase. The plan for the first phase, covering the years 1960–2, was formulated in 1959 and approved by the cabinet in January 1960. Three months later, the Rhee government was overthrown and the plan was abandoned.[52]

The Chang Myon regime that came into power immediately following the Student Revolution in April 1960 prepared a draft

TABLE 2.17 Key indicators of the South Korean economy, 1954–1961

	1954	1955	1956	1957	1958	1959	1960	1961	1954–61 (average)
GNP growth rate (%)[a]	5.1	4.5	−1.4	7.6	5.5	3.8	1.1	5.6	4.0
Per capita GNP (US$)	70	65	66	74	80	81	79	82	—
Investment ratio (%)	11.6	11.9	9.5	15.3	13.0	10.7	10.9	13.1	12.0
Domestic savings ratio (%)	6.4	4.9	−1.3	5.5	5.0	3.9	1.4	3.9	3.7
Wholesale price index (1985 = 100)	1.53	2.76	3.65	4.22	3.96	4.09	4.50	5.10	3.7
Consumer price rise[b] in Seoul City (%)	27.3	68.0	23.5	23.1	−3.1	4.3	8.3	8.0	21.2
% of GNP									
Agriculture, Forestry and Fishery	40.3	44.8	47.2	45.2	41.2	34.7	36.8	40.2	
Mining and Manufacturing	12.4	12.2	12.5	12.5	14.1	15.8	15.7	15.2	
(Manufacturing sector only)	(11.5)	(11.2)	(11.3)	(11.0)	(12.6)	(14.0)	(13.6)	(13.4)	
Social services & others	47.3	43.0	40.3	42.3	44.7	49.5	47.5	44.6	
Exports (US$ million)	24	18	25	22	17	20	33	41	200
Imports (US$ million c.i.f.)	243	341	386	442	378	304	344	316	2754
Foreign aid (US$ million)	153.9	236.7	326.7	382.9	321.3	222.2	256.1	199.2	
Exchange rate (W/US$)[c]	18.00	50.00	50.00	50.00	50.00	50.00	65.00	130.0	

[a] At 1985 constant prices.
[b] For 1954–61, consumer inflation rate; after 1961, consumer price index.
[c] At end of year.

Sources: Economic Planning Board (EPB), *Handbook of Korean Economy* (1977), 2–5; Bank of Korea, *Principal Economic and Social Indicators* (1945–88).

for a new five-year plan, which was completed just before the military coup of May 1961. The plan suffered the same fate as the previous one, in that it was not accepted by the new government that assumed power after the coup. It did, however, provide the basis for a third planning attempt, which was finally approved in late 1961 as the first Five-Year Economic Plan.[53]

(a) The first and second Five-Year Plans: 1962–1966 and 1967–1971   After the military coup of 16 May 1961, the newly established Supreme Council for National Reconstruction led by General Park Chung-Hee assumed all legislative, executive, and judicial powers. It set up new guidelines for implementation of economic development plans aimed at overcoming the incessant vicious circle of poverty. It transformed the then Economic Development Council into the Economic Planning Board (EPB), combining the planning, budgeting, and foreign assistance administrative functions. The EPB was assigned as its priority task the revision of the draft Five-Year Plan (1962–6) to conform to some new guidelines laid down by the Supreme Council.[54] This was then reviewed by various Korean advisory groups during the last quarter of 1961 and finally approved by the Supreme Council.

The second Five-Year Economic Development Plan was prepared in 1965 and 1966 and was approved by the President in August 1966. This basic programme has since been supplemented and modified by annual plans, called Overall Resource Budgets (ORBs), prepared each year since 1967.

The gist of the first and second Five-Year Plans was basically to build an industrial base for both import substitution and export-oriented industrialization, principally through the following efforts and measures: (1) facilitating the construction of overhead capital; (2) ensuring self-sufficiency in food-grain production; (3) creating and expanding the cement, chemical–fertilizer, synthetic fibre and other import-substitution industries; (4) overcoming the limitation of narrow domestic markets; (5) attracting sufficient foreign capital to overcome the vicious circle of poverty resulting from a shortage of domestic savings and investment; and (6) laying the foundation for sustained export growth to increase employment. As W. Arthur Lewis adequately pointed out, there is no single unifying theme to a discussion of development strategies: instead, there must be a judgement of which issues are important in a particular setting.[55]

The single most important objective of the first Five-Year Plan was the promotion of industrialization and modernization of the economy through the efficient allocation of resources and a balanced industrial structure. To achieve this objective, the planners set strategy to concentrate on certain key types of specific project, including infrastructure and social overhead capital and the maximization of labour-intensive methods of production. The important question was whether individual projects were to be carried out by the government or the private sector or both, and how these projects would be financed.

The indicative planning undertaken in the first Five-Year Plan involved the establishment of sectoral targets which were not compulsory for the private sector. However, in some key strategic industries and in the social overhead capital sector, the plan did not intend that the market should be entirely responsible for the allocation of resources and investment. Instead, the government either intervened directly in the detailed allocation of resources or else influenced the market through credit rationing and regulations. Termed 'the guided capitalist system', this practice was later to lead to some serious problems. Private entrepreneurs tended to assume that they would automatically receive financing once they were authorized to carry out a specific project listed in the plan. Because the government supplied them with preferential treatment and helped them to obtain financing at a lower rate of interest, some entrepreneurs recklessly expanded their business complexes under this government umbrella, without carefully considering the marketability of their products and the marginal rate of return. This practice continued until the mid-1980s, when it came under severe criticism as the partial cause of ailing firms and business instability.

During the first Five-Year Plan (FFYP) period, the total amount of investment spent was W321.5 billion, of which 48.8% was allocated to the electricity, transport (and road construction), communication, and housing sectors, 34% to the mining and manufacturing sector, and 17.2% to the primary industry sector. Government took charge of 34.8% of this investment; the remaining 65.2% was carried out by the private sector out of total investment resources, of which 55.6% was financed by the government sector and 44.4% by the private sector. In fact, the government contributed a large portion of gross domestic capital formation in the early phase of several plans because of the relatively weak capability of the private sector to accumulate capital. Gross domestic investment consisted of 42.1% of national savings (W135.5 billion) and 57.9% of foreign savings (W186.0 billion, or $1.43 billion) during the FFYP period.

The first Five-Year Plan was carried out with substantial revisions which were often made arbitrarily. However, its performance was remarkable, with a rapid growth in GNP, improvements in industrial structure, and, more importantly, the awakening of the people's vision for an improved standard of living.

GNP grew at an average rate of 7.9% per year, 0.8 percentage points higher than the original goal of 7.1% per year. Thus, per capita income reached $125 in the target year, which represented a 43.7% rise from the 1962 level of $87.

Secondary industry (the mining and manufacturing sectors) grew at 15.0% per annum, leading overall growth, while primary and tertiary industries grew at 5.3% and 8.1%, respectively. As a result, the share of the secondary industry in the composition of industrial output increased from 15.7% in 1960 to 20.5% in 1966, while that of the primary industry dropped from 36.8% to 34.8%, and that of the tertiary industry from 47.5% to 44.7%. However, FFYP's performance was generally short of its plan target, as can be seen in Table 2.18, in spite of the overperformance in GNP growth rate, largely as a result of disappointing savings growth, an inability to raise government revenues, and poor farm harvests in 1962–3, not to mention the maladministration of government officials.

Circumstances at the start of the second Five-Year Plan (1967–71) were very different from those at the start of the first. Growth had accelerated and business had improved during the first Five-Year Plan period. In particular, the Korean currency (*won*) devaluation in early 1964 began to have a favourable effect on export competitiveness, while an interest rate reform in autumn 1965 funnelled savings to banks from the curb market.[56]

The second Five-Year Plan (SFYP) was a success in that plan targets were greatly exceeded. The plan called for GNP to rise 50% from 1965 to 1971 and exports to increase by 148%: the actual increases (in 1975 prices) were 79% and 417%, respectively.

In addition to both exchange rate and interest rate overhaul, the government carried out price stabilization and tax reform

TABLE 2.18 Projection v. performance in each of South Korea's Five-Year Development Plans: annual averages[a]

| | First Plan (1962–6) | | Second Plan (1967–71) | | Third Plan (1972–6) | | Fourth Plan (1977–81) | | Fifth Plan (1982–6) | | Sixth Plan (1987–91) | | Seventh Plan (1992–6) |
	(A)	(B)	(A)	(B)	(A)	(B)	(A)	(B)	(A)	(B)	(A)	(B)	(A)
GNP growth rate (%)	7.1	7.8	7.0	9.6	8.6	9.7	9.2	5.8	7.5	8.6	11.1	10.0	7.5
Industrial structure (%)													
Primary	34.8	34.8	34.0	26.8	22.4	23.5	18.5	15.8	12.2	12.8	11.4	9.1	7.3
Secondary	26.1	20.5	26.8	22.2	27.9	28.4	40.9	30.7	31.0	30.1	31.6	29.7	30.6
Commodity exports ($ million)	140	250	600	1100	3500	17 800	20 200	20 700	35 700	33 600	47 200	61 400	106 300
Commodity imports ($ million)	490	680	900	2200	2800	8400	18 900	24 300	35 100	29 300	43 600	61 100	111 400
Gross investment ratio (%)	22.7	21.6	19.9	25.1	24.9	25.6	26.0	30.3	29.5	29.5	30.7	36.3	36.4
Domestic savings ratio (%)	13.0	11.8	14.4	14.6	21.5	23.9	23.9	20.5	21.7	32.5	33.2	34.0	35.5
Unemployment rate (%)	14.8	7.1	5.0	4.5	4.0	3.9	3.8	4.5	3.8	3.8	3.7	2.6	2.5

[a] (A), Plan projection; (B), Performance.

Source: Economic Planning Board (South Korea).

programmes to control inflation and raise the ratio of tax revenue to GNP starting in 1966. Trade promotion and tariff reform programmes followed in 1967. The stabilization programme and a set of export promotion measures worked to increase output and productivity, particularly in export production. With a rapid growth of the economy, important structural changes took place. The mining and manufacturing sectors increased their share of GNP from 16% in 1962 to 22% in 1971, while the agriculture sector dropped from 37% to 27%. Commodity exports, which amounted to only about $55 million in 1962, grew at an annual average of 40% to a total of $1.10 billion in 1971 on a customs clearance basis. Commodity imports also expanded markedly, from $422 million to $2.39 billion.

Thus, the South Korean economy was fully committed to the outward-looking development strategy set by Park Chung-Hee with his commitment to 'nation-building by export'. As the economy continued to progress, but with strong reliance on foreign savings, in the late 1960s and early 1970s, many Koreans came to wonder whether too much foreign dependency would endanger national independence. Since South Korea did not have a credible debt servicing capacity as measured by export earnings until around 1972,[57] such criticisms seemed well grounded. But the outward-looking development strategy has subsequently proved to be the right policy option, as the country's balance of payments has steadily improved ever since.

(b) The third and fourth Five-Year Plans The third Five-Year Plan (1972–6) and the fourth Five-Year Plan (1977–81) included the modernization of rural areas in addition to a continued emphasis on export expansion and the development of heavy and chemical industries. The third Five-Year Plan (TFYP), in particular, called for the dynamic development of the rural economy and balanced regional development. Paul W. Kuznets notes:[58]

What the plan does not show is the erosion of rural support for the government in the election of 1971 and the subsequent establishment of *Samaul Undong* (new community movement) to improve rural life, the phasing out of American PL 480 (food grains) assistance, or the massive rural–urban migration of the late 1960s, evidence that rural areas were not fully sharing the benefits of development. Nor does the plan mention the Nixon Shock when President Nixon ended the dollar's convertibility to gold, or the one-third reduction in the US troop level in Korea, both of which occurred in 1971.

In formulating the third Five-Year Plan, the government put much emphasis on both 'Samaul Undong' (the 'New Village Movement') and the heavy and chemical industries. The Park Chung-Hee government was greatly shocked by the decreased support from rural voters in the 1971 elections,[59] and also by the sudden US *rapprochement* with China in 1972. The ruling party leaders badly needed to wheedle the support of rural voters in order to keep in power, and they recognized the inevitable need to provide South Korea with its own means of defence against external threats. All these concerns affected the TFYP by increasing the focus on the rural sector and on the heavy and chemical industries. In addition, this emphasis would serve to offset growing world protectionism by raising the value-added component of Korea's exports.

Despite the worldwide chaos caused by the first oil shock in 1973, the third Five-Year Plan was as successful as the second, and its even more ambitious plan targets were exceeded. The oil shock had the expected disruptive effects on the South Korean terms of trade and balance of payments. The country's current account deficit jumped to more than $2 billion each year in 1974–5, or 11% of GNP. Concern for this deterioration in the balance of payments is evident in the fourth Five-Year Plan (1977–81) targets, which included self-reliance in investment financing and, with growing exports of construction services to the Middle East, a return to balance-of-payments equilibrium by 1981.

Nevertheless, the fourth Five-Year Plan goals were not met. Actual GNP growth averaged 5.8% a year in 1977–81 whereas the plan target was 9.2%. The current payments account for 1981 was $4.7 billion in deficit (valued on FOB basis), and South Korea was further from rather than closer to self-reliance in investment financing. One reason why the plan targets were not met was the second oil shock in early 1979. In South Korea this shock was followed not only by political disruption with President Park's assassination on 26 October 1979, but also by a disastrous harvest in 1980.

The jump in oil prices was further amplified by the perverse macroeconomic policies that had developed along with the energy-intensive heavy and chemical industries. Also, overinvestment in the heavy and chemical industries gave rise to several additional adverse effects on the economy, including insufficient investment in light and consumer industries, distortions in the capital market owing to state interference, and high wage increases, leading to price rises. The stabilization programme which was initiated in early 1979 could not bear fruit because of both domestic and foreign man-made and natural failures. High inflation hit the economy in 1980, when GNP dropped for the first time since 1956.

(c) The fifth and sixth Five-Year Plans The dramatic emergence of the so-called Fifth Republic government under General Chun Doo-Hwan amid political turmoil in 1979–80 enabled economic planners to wipe the slate clean in drawing up the fifth Five-Year Plan (1982–6). This plan called for price stabilization and for the privatization (liberalization) of economic activities, while the basic policies pursuing regional balance and skill-intensive machinery and electronics manufacture would remain intact. The government was deeply committed to promoting free trade, liberalizing financial policy, and dismantling and reducing state regulation and intervention in the market mechanism. The fifth Five-Year Plan also called for the establishment of a comprehensive social security system in order to reduce inequalities among income classes and between regions.

Although this plan focused on institutional improvements to resolve structural problems rather than on setting goals in quantitative terms, it did set goals, and the performance was good. The actual GNP growth rate in 1982–6 was 8.6%, exceeding the revised fifth Five-Year Plan's target rate 8.0%, while the current account surplus in 1986 ($4.6 billion) was well above the plan's $0.4 billion target. This last discrepancy had been attributed to 'three blessings': the collapse of world oil prices, a decline in international interest rates, and the US dollar devaluation (or Japanese yen appreciation) all worked in South Korea's favour. From 1986 to 1990, the economy enjoyed the blessings of a current account surplus for the first time, but it was experiencing deficits again (temporarily) in the first half of 1990.

South Korea's current plan, the sixth Five-Year Plan (1987–91), calls for an annual inflation rate over 1985–91 of 3.1%, an increase in real GNP of 7.1% a year, and a current account surplus that will average $5 billion a year. Actually, GNP grew by 13.0% in

1987, 12.4% in 1988, 6.7% in 1989, and 9.0% in 1990, demonstrating that the South Korean economy remains robust amid many labour disputes in the course of the democratization process. With 1990 as a watershed, the development pattern of South Korea is shifting from an outward-looking to an inward-looking one, as the planners intend to reduce the country's heavy dependence for economic growth on external trade.

Economic management: plan formulation and implementation

Early economic planning in South Korea was done mainly by foreigners—the Nathan team and others associated with the UN and US assistance programmes—because there were very few Koreans with any experience in this area.[60] But from 1961 onwards, planning has been the province of the Economic Planning Board (EPB), a super-ministry established in 1961. The EPB is responsible for planning, budgeting, and evaluation as well as statistical work, and is headed by the deputy prime minister. The Bureau of Economic Planning (BEP) within the EPB carries out the overall planning in co-operation with each ministry's Office of Planning and Management, which is a member of the *ad hoc* committees that prepare the sectoral plans. The *ad hoc* committees or work-groups include ministry officials, Bank of Korea and other development bank representatives, private individuals, and experts.

The planning cycle begins with the EPB drafting guidelines, which usually include an assessment of the external economic situation, macroeconomic targets, and policy directives. The guidelines are submitted to a vice-ministerial committee for inter-ministerial co-ordination and to ministerial committees for approval, after which they are presented at a formal cabinet meeting and to the president. They are then forwarded to all government ministries, so that each ministry can prepare sectoral plans on the basis of the guidelines. Sectoral plans are adjusted and co-ordinated after being submitted to the BEP before being integrated to form the draft national plan, at which point they are sent to the Vice-Ministerial Committee of Plan Co-ordination for review and approval. The EPB then updates the plan, submits it to the cabinet and the president for final approval, and lastly writes and publishes the final version.[61]

The plans are implemented through annual plans and government budgets. Annual planning began with the second Five-Year Plan's Overall Resource Budgets (ORBs) which were designed to review and evaluate performance, revise policy and project implementation, and provide guidelines for preparing the government budget. The ORB was changed to Economic Management Plan (EMP) from the fourth Five-Year Plan period.

Based on macroeconomic indicators prepared by the BEP for the coming year, each ministry and independent central government agency prepares its own EMP and budget. These are consolidated into a final budget which is prepared by the Ministry of Finance (MOF) and must be approved by the National Assembly.

The EMPs are annual rolling plans and are subject to revision as necessary during the year. They keep government budgets consistent with Five-Year Plan goals and they constitute the overall source of information of government policies and programmes.[62] The ORBs and EMPs have made it possible for the government to employ a flexible and pragmatic policy approach over time rather than sticking obstinately to a pre-fixed intermediate or long-term plan. In other words, short-term flexible macroeconomic policy has been and is being adequately utilized in South Korea. Although the government formulates economic plans and retains its influence, the core of economic operation is largely in the hands of the private sector. Of course, the private sector has responded to government plan directives, because a combination of incentives and preferential credit rationing measures makes it profitable to comply.

2.3 SUMMARY

Any economic policy or plan aims to promote economic growth and development, which constitutes one standard by which to judge the success of a country's economy. Economic growth and development has been a high-priority goal for both North and South Korean regimes. Beginning with a common cultural background and a shared history, but proceeding with differing ideological systems as well as different endowments of natural resources, the two Koreas have chosen radically different economic policies and plans by which to develop their economies and promote the welfare of their people.

The important aspects of the North Korean system are state ownership of the means of production, a command economy, and policies giving priority to heavy industry over light industry. Another feature has been the very autarkic, inward-looking, and ideologically bounded nature of the economic plans. Above all,

North Korean planning appears to vary in some important respects from that of other communist economies. First, unlike other communist countries (and unlike its rival South Korea), North Korea has formulated its economic plans not on a five-year basis, but typically for a six- or seven-year period. Presumably, this is because the planners want to avoid having their plans integrated into those of other communist countries. North Korea has so far resisted joining the Council for Mutual Economic Assistance (CMEA), which until recently has closely co-ordinated the economic plans of communist countries. Perhaps the North Korean leaders also want to discourage a direct comparison of its economic targets and strategies with its rival in the South.

Second, North Korea is more flexible about the plan period and less wedded to the norm of the annual plan. Often it will make unannounced changes, adjusting targets as well as plan periods as inevitable situations occur. Generally, North Korea appears inclined to place greater emphasis on specific micro projects, under the slogan of 'speed battles', than on overall macro targets. The vagaries of the economy and other circumstances often lead to objectives being shifted from one plan period to the next, and sometimes the focus of a plan shifts partway through. North Korea's stable leadership knows what it wants but has to be flexible in achieving its goals.[63]

The North's economic planning system was overhauled in 1981. The Central People's Committee, consisting of ministries and commissions, retained responsibility for planning and technical guidance overall, but responsibility for supervising factories and fulfilling quotas was entrusted to a new provincial economic guidance committee directly under the provincial party organizations. Presumably this has made provincial economic achievements somewhat more important, but whether this has been transformed into political clout at the upper reaches of the party is not known to outsiders.[64]

South Korea, on the other hand, operates with a mixed economic system; although principally it embraces capitalism and private enterprise, there is government indicative planning, guidance, and regulation. It has been a market economy, but until recently one subject to substantial government management and intervention. South Korea initially gave priority to labour-intensive light industry, waiting until later to develop capital- and skill-intensive heavy and chemical industries. South Korean economic policy was outward-looking and depended heavily on external sources of raw materials, markets, capital, and technology.

The North Korean economy grew faster than that of South Korea during the first decade after the Korean War, but since the late 1960s South Korea has achieved tremendous economic growth, exceeding that of North Korea in the 1970s and 1980s.

The South Korean economy has wrought a drastic industrial structure change on the country, as its balance of trade improved from chronic deficits to surpluses since 1986.

A comparison of the macroeconomic performances of North Korea and South Korea shall be attempted in the next chapter.

NOTES TO CHAPTER 2

1. Vaclav Holesovsky, *Economic Systems: Analysis and Comparison* (New York: McGraw-Hill, 1977), 31.
2. Kim Il-Sung began to attack the formalism and doctrinism within the party in April 1955 before he first used the word '*chuche*'. This reflected a sort of power struggle going on within North Korea's ruling party. See Kim Il-Sung's *Selective Work Book* no. 4 (1960), 230–7.
3. *Chosen Chung-Ang Nyun-Gam* (North Korea's Central News Annual Report) (1963), 157–62.
4. *Rodong Sinmun* (North Korea's official newspaper), 13 January 1990, 1, 3; *Rodong Sinmun*, 'Party Organizations Should Vigorously Conduct the *Chuche* Idea Indoctrination by Linking It with the Implementation of the Revolutionary Tasks' (editorial), 24 February 1990.
5. Pyongyang, KCNA in English 04.58 GMT, 26 March 1990; see FBIS-EAS-90-058, 26 March 1990, 27–8. See also *Rodong Sinmun*, 'The *Chuche* Idea is a Revolutionary Ideology which Has Given a Profound Elucidation on Independence', special article, 25 March 1990.
6. To help readers understand the earlier socialization process of North Korea, the chronological order of reforms is given:

5 March 1946	Promulgation of law on land reform
10 August 1946	Promulgation of law on nationalization of the major means of production
22 December 1947	Promulgation of law on nationalization of mineral resources, forests, and water basins
April 1954	Initial organization of co-operative farms
August 1958	Completion of co-operatization of farmers, handicraftsmen, merchants, and industrialists
October–December 1958	Reorganization of co-operatives on scale of village (*Ri*) unit

(*Source*: South Korea's National Unification Board, *Survey of North Korean Economy* (Seoul, 1989), 10.)
7. For the development of North Korea's industrial structure, see Chs. 3 and 6.

8. Here the source of social accumulation is defined as part of the social gross products or net social income. Broadly speaking, it is defined to include the sum of social net income (m) and bonus pond (V) paid to workers to compensate for their consumption (costs) in production process. See Ahn Kwang-Zup, *Socialistic Accumulation in the People's Economy* (Pyongyang: Academy of Social Sciences, 1964), 73–85.

9. Labour productivity's contribution to social gross output growth accounted for about 68% during the three-year reconstruction plan periods after the Korean War. But for the first Seven-Year Plan period (1961–7) labour productivity growth accounted for about 75–80% of total output growth in North Korea. Meanwhile, the growth rate of labour productivity was estimated to be 370% in 1960 over 1953, while money wage income rose by 370%, but the consumer price index dropped by 51% over the same period. This shows that real-wage income growth exceeded the growth of labour productivity during those years after the war in North Korea. See Ahn, *Socialistic Accumulation*, 86 and 115.

10. The concept of national income used in the communist economy differs from that of GNP in the capitalist system. National income in the communist system does not include value added in most service sectors or depreciation costs. But it does include the transaction revenues, which are equivalent to the differences between wholesale prices and retail prices in the transaction of some consumer goods and some services. The value of social products in the communist economy entails some double accounting in the production process, because it is estimated by adding up all value added in every stage of the production process. For a more detailed explanation of the various income concepts of North Korea, see Ch. 3.

11. In principle, the total value of output achieved in excess of the state plan target rate (100%) is distributed to workers. The basic straight wage paid for work done within the state targeted goal (100%) changes in proportion to the accomplishment rate of the plan target. However, the reward (bonus) payment increases in accordance with the rate of overfulfilment. Note that the state basic fulfilment rate is based on 100%, while the overfulfilment rate is based on 1%, since it is applied to every portion beyond basic plan goal (100%). For example, if the actual fulfilment rate is 110%, it means 10% overfulfilment as compared with the targeted 100% plan. But it is equivalent to just 10 times or 1000% if measured in terms of every 1% of overfulfilment. See Ahn, *Socialist Accumulation*, 108, for further details of this complicated method for computing the labour productivity growth rate and reward rate used in North Korea.

12. Philip Rudolph, *North Korea's Political and Economic Structure* (New York: Institute of Pacific Relations, 1959), 61–4. See also Yoon T. Kuark, 'North Korea's Industrial Development During the Post-War Period,' in Robert A. Scalapino (ed.), *North Korea Today* (New York/London: Frederick A. Praeger, 1963), 51, n. 1.

13. Bruce Cummings, *The Origins of the Korean War: Liberation and the Emergence of Separate Regimes 1945–1947* (Princeton, NJ: Princeton University Press, 1981), ch. 11. Evidence shows that during the post-war years shared foreign and domestic experience tended to produce factional and hegemonic political groupings in the North. But Kim Il-Sung won hegemony in the initial stages and exercised considerable autonomy. For this, see Kim Il-Sung's 1955 speech 'On Exterminating Dogmatism and Formalism and Establishing Independence in Ideological Work' (Pyongyang: KWP Press 1960), 12. The following passage from his speech is self-explanatory: 'Although some people say that the Soviet way is best or that the Chinese way is best, have we not now reached the point where we can construct our own way?'

14. Jon Halliday, 'The Economics of North and South Korea', in John Sullivan and Robert Foss (eds.), *Two Koreas—One Future?* (University Press of America, 1989), 22.
15. See n. 6 above.
16. Frederica M. Bunge (ed.), *North Korea: A Country Study* (Washington, DC: US Government Printing Office, 1981), 139–46, and Joseph S. Chung, *The North Korean Economy: Structure and Development* (Stanford, Cal.: Hoover Institution Press, 1974), 5–43.
17. *Chosen Chung-Ang Nyun-Gam*, (1961), 322.
18. Ibid.; see 322–32, tables 1–30.
19. For economic aid received by North Korea, see below.
20. The *won* figures (in terms of 1950 prices) for this Three-Year Plan period are in old *won* of North Korea. On 17 February 1959, the second currency reform (the first was in 1947) was made, by which 1 new *won* was exchanged for 100 old *won*. (See *Kyongje Konsul* ('Economy Construction') (Pyongyang), no. 2, (February 1959), 2–3.) After this currency reform, the exchange rate between the British pound and the North Korean *won* was quoted unofficially to be £1 : W7.2. Therefore, the US dollar may be valued at W2.5; the Japanese yen at ¥150 to W1. See *Kita-chosen no Kiroku* ('Records of North Korea') (in Japanese) (Tokyo: Shindokusho-sha, 1960), 132–3; and Scalapino, *North Korea Today*, 52, n. 4. But the official rate fluctuated in the range of W0.90–1.05 per US dollar during the 1946–89 period.
21. 'Productive construction' in North Korea includes: (*a*) industrial construction, (*b*) agricultural construction, (*c*) transportation and communications construction, and (*d*) construction for commerce and social services; while 'non-productive construction' includes: (*a*) educational and cultural construction, (*b*) scientific research, (*c*) housing, and (*d*) public facilities like parks, museums, public restrooms, etc.
22. In the early stages of organizing farm co-operatives, about 74% of farm families owned only 1–2 cheongbo of small lands. Therefore the co-operative size was also small because each cooperative was nothing but a common land use unit in nature. But the average co-operative scale has been enlarged since 1964. See Kim Il-Sung's *Selected Papers*, no. 8 (Pyongyang: Chosen Workers' Party, 1985), 393–495.
23. *Chosen Chungang Nyungam* ('Korean Central Yearbook'), *1961* (Pyongyang: North Korea Central News Agency), 323, table 3.
24. *Chosen Chungang Nyungam* (Korean Central Yearbook), *1958* (Pyongyang: North Korea Central News Agency), 102–4. But South Korean analysis of North Korea's first Five-Year Plan indicates that the Plan achieved only 77% in industry and 90% in agriculture out of plan targets. See Research Centre for North Korea, *North Korea Survey* (Seoul, 1977), 561.
25. Hideo Akimoto, 'On the Economy' and 'Topics on Heavy Industry', *Record of North Korea* (Tokyo, 1960), 75–118. Mr Akimoto of the *Tokyo Yomiuri Shinbun* (newspaper) gave an eye-witness account of the miraculous story of how the Five-Year Plan was achieved within two and a half years. In sum, he cited: (1) worker enthusiasm, (2) extensive mobilization of labour and resources, (3) increase in labour productivity, (4) an effective incentive system, (5) increasing industrial technology, and (6) material and technical aid from the Soviet Union (Rb1000 million roubles), China (¥800 million), Czechoslovakia, East Germany (capital equipment), etc., as reasons for this achievement. Also refer to Scalapino, *North Korea Today*, 53–4, n. 8.
26. For relative composition of capital formation (investment) in heavy industry *v.* light industry during the plan period (1950–60), see Tables 2.2 and 2.4.

27. See n. 20 above. The economic effects of this monetary reform are not known because of the lack of statistics.
28. This is related to the Chŏngsan-ri method taught by Kim Il-Sung. It emphasizes doing farm work on a scientific and technological basis. See Kim Il-Sung's classic work, *Theses on the Socialist Rural Question in Our Country* (Pyongyang, 1964).
29. Since 1965, North Korea has published only piecemeal economic statistics or data on an individual commodity basis. The release of this sort of datum is neither consistent nor regular in types and items. Therefore, depending on who collects and organizes the data, there appear to be wide variance among different users or estimates.
30. Teruo Komaki, 'Current Status and Prospects of the North Korean Economy', in M. Okonugi (ed.), *North Korea at the Crossroads* (Tokyo: Japan Institute of International Affairs, 1988), 58.
31. Ibid. 58–9.
32. DPRK economic officials might be well aware of this fact, but, because of a vain, self-centred obsession with political creed and slogan, it may be difficult for them to take a flexible approach. This is an obstacle not only to an economic opening towards the outside world, but also to any economic reform within the economy.
33. Foreign Broadcasting Information Service (FBIS), Daily Report: East Asia, 1 June 1990, 20.
34. See Ha-Kwang Kim, 'Pride in Having Superior Economic Management System which Gives Powerful Impetus to Consolidation and Development of Socialist System', *Rodong Sinmun*, 20 December 1989.
35. The 'Taean Work System' was proposed by Kim Il-Sung when he visited an electric manufacturing plant in December 1961. It proposes that the party committee shall be responsible for production lines by its members working together with workers. Along with 'Chongsan-ri spirit and method' suggested by him in February 1960, this system provides the main economic management guidelines in North Korea.
36. 'Transaction revenue' is a sort of sales tax applied to both consumer goods and intermediate goods. In North Korea it is regarded as a part of net social income rather than being an indirect (sales) tax. It constitutes the largest element of budget revenue in North Korea. For details of this definition, see Ch. 3.
37. This poem, originally written in Korean, was translated by Peter Kim.
38. Whenever I look back to those days of the Korean War, I cannot help reminding myself of the biblical story of the prodigal son in need 'who longed to fill his stomach with the pods that the pigs were eating, but no one gave him anything' (Luke 15: 14–16).
39. Edward S. Mason, Mahn-Je Kim, Dwight H. Perkins, Kwang-Suk Kim, and David C. Cole, *The Economic and Social Modernization of the Republic of Korea* (Cambridge, Mass.: Harvard University Press, 1980), 93.
40. Ralph N. Clough, *Embattled Korea: The Rivalry for International Support* (Boulder, Colo.: Westview Press, 1987), 68. See also Charles R. Frank, Jr, Kwang-Suk Kim, and Larry E. Westphal, *Foreign Trade Regimes and Economic Development: South Korea* (New York: Columbia University Press, 1975), 8–9. Note here that the industrial production of South Korea in 1953 is compared with the industrial output of the whole of Korea in 1940.
41. These dollar values of exports and imports were estimated by applying the average official exchange rate of W0.68 to $1 in 1949 to the current-price *won* values of exports and imports. To find commodity exports and imports of South Korea during the period 1945–53, see table 2-3 in Frank *et al.*, *Foreign Trade Regimes*, 10.

42. Ibid.
43. In 1946 and 1947, barter trade took place at South Korean ports between foreign merchants, mainly Chinese from Hong Kong and Macao, and Korean exporters. The Koreans, who lacked both experience and capital, acted mainly as brokers for the foreign merchants. See Frank *et al.*, *Foreign Trade Regimes*, 10.
44. This part was partially quoted from an unpublished working paper on 'Land Reform in Korea' which I prepared when working for Dr Fletcher Riggs and Dr Robert B. Morrow at the Rural Development Division, US AID/Seoul, Korea, 1969–71. See also Robert S. Morrow and Kenneth H. Sherper, *Land Reform in South Korea* (Seoul: US AID/Korea, 1970).
45. One cheongbo is equivalent to 2.45 acres, and 1 acre is equal to 0.4046856 ha. Therefore, to convert from cheongbo to hectare, multiply first by a factor of 2.45 and then by 0.4046856.
46. Nathan and Associates (1954) estimated South Korea's GNP for 1952 and 1953 at $1384 million and $1721 million. See Frank *et al.*, *Foreign Trade Regimes*, 11.
47. Public Law 480 (PL 480) was the US Agricultural Trade Development and Assistance Act, enacted in July 1954. This was aimed at selling US farm products abroad and using the money to aid the economic development of less developed countries.
48. Kwang-Suk Kim and Michael Roemer, *Growth and Structural Transformation* (Cambridge, Mass.: Harvard University Press, 1979), 41–2.
49. The comparisons of macroeconomies between North and South Korea are attempted in a later chapter. For details, see Ch. 3. There had been some economic growth (approximately 3.7% per year) in the Rhee government period, largely as a result of US economic assistance, and agricultural production was again at prewar levels.
50. The legacy of the Rhee government was not enviable: autocratic leadership; a political process that could stand little careful scrutiny; only a modicum of international credibility; lack of economic management; a record of extreme corruption; and pent-up frustrations that resulted in the fall of the Rhee regime in spring 1960, through a students' uprising.
51. Park Chung-Hee, *To Build a Nation* (Washington, DC: Acropolis Books, 1971), 16.
52. Irma Adelman (ed.), *Practical Approaches to Development Planning: Korea's Second Five-Year Plan* (Baltimore: Johns Hopkins University Press, 1969), 12.
53. Ibid. 12.
54. Arthur D. Little, Inc., *Economic Development Planning in Korea: Report of the Arthur D. Little Reconnaissance Survey* (Seoul, 1962).
55. W. Arthur Lewis, *Development Planning: The Essentials of Economic Policy* (New York: Praeger, 1966), 26; and Adelman, *Development Planning*, 21.
56. In order to achieve the basic goal of export-oriented industrialization and high growth, a package of policy reforms took effect between 1964 and 1967. The *won* was devalued by almost 100%, from W130 to W256 per US dollar in May 1964, and a unitary floating exchange rate system was adopted in March 1965. Following the exchange rate reform, the government almost doubled the interest rates on bank deposits and loans in September 1965 in order to increase voluntary private savings, and as a result deposits doubled every year for the next three years. Also, any unproductive bank credit was discouraged by the government.
57. The debt service ratio (defined as the percentage of foreign loans' principal plus interest to foreign exchange earnings) was 5.4% in 1967 and 1968, 8.6% in 1968, 18.5% in 1969, 21.0% in 1970, 19.7% in 1971, 18.4% in 1972, 14.3%

in 1973, 11.4% in 1974, 12.5% in 1975, 12.1% in 1976, 11.9% in 1977, 13.9% in 1978, 16.3% in 1979, 18.8% in 1980, 20.7% in 1981, 21.0% in 1982, and 19.2% in 1983.

58. Paul W. Kuznets, 'Planning in Korea', paper presented at a conference on 'Indicative Planning', Brookings Institution, Washington, DC (April 1990), 5–6.

59. The presidential election was held on 27 April 1971. Park Chung-Hee had four opponents, of whom New Democratic Party candidate Kim Dae-Jung was the one of consequence. There were 15 552 236 people eligible to vote, and of these 11 923 218 cast valid ballots. The vote for Park Chung-Hee was 6 343 828 and for Kim Dae-Jung, 5 395 900. In the vote for the National Assembly four weeks later, Park's Democratic Republican Party won 113 out of 204 seats, the NDP won 89, and 2 went to splinter parties. The results can be compared with those of an earlier election held on 15 October 1963. Then, votes cast totalled 11 036 175. Of these Park won 4 702 640, against Yun Po-Sun's 4 546 614. The remaining votes were shared among three also-rans. In elections for the National Assembly on the following 26 November, Park's newborn DRP won 110 of the available 175 seats, with the remaining 65 divided between four opposition parties.

60. Adelman, *Practical Approaches*, 13.

61. Kuznets, 'Planning in Korea', 12–13.

62. Ibid. 13–14.

63. John Merrill, 'North Korea's Halting Efforts at Economic Reform', paper presented to the Fourth Conference on North Korea, Seoul, 7–11 August 1989.

64. Ibid. 8.

3

An Economic Comparison of North and South

No matter how difficult it is, we must come to terms with the fact that there was an alternative, another route that did not entail such tragic casualties and useless waste of resources, such suppression of incentives to work, such a decline in moral certainties. In the end, the success of the economic reform emerging today depends on how profoundly we have learned the lessons of our own history.

'The Turning Point' by Nikolai Shmelev and Vladimir Popov[1]

3.1 A FEW WORDS ABOUT ECONOMIC STATISTICS

Even if we set aside such differences between North and South Korea as the initial endowments of natural resources, industrial structure, the ownership pattern of the means of production, and economic planning, there are several hazards in attempting a comparative analysis. First, in comparing economies of two or more countries, a common measuring stick is usually necessary. Gross national product (GNP) and per capita income are two of the best measures known for such purposes. However, difficulties arise when economic comparison is attempted between a market economy and a centrally planned economy (CPE). The Democratic People's Republic of Korea (DPRK—North Korea) has kept a command, or centrally planned, economy, while the Republic of Korea (ROK—South Korea) has a market economy. As in other CPE countries, North Korea's national accounts are based on the concept of *net material product* (NMP), which differs from the gross national product (GNP) concept used in market economies.

In capitalist market economies, GNP (at market prices) is the total value of all final products and services produced during a given period of time. National income (NI) is the total of all incomes received by all factors of production. The national income accounts of CPEs, however, record the productive activity carried out in their country, rather than the income received by their residents. They are based on the *material product system*, whose most important aggregate is NMP, and its derivation is fairly standardized. The NMP comprehensively covers value added in the 'material' sectors of production. The sum of the outputs of all separately enumerated production units (material products: MP) multiplied by relevant market prices of outputs is called the *gross output value of social production* (GVSP), which includes also the values of intermediate products. The values of intermediate products are simply the material costs for consecutive production units, so in the GVSP concept the values of products are counted more than once. This is not a problem, because for a CPE national income represents the sum of net product (value added) of all separately enumerated branches of the economy. Therefore NI equals GVSP minus all material costs including depreciation of capital, of which NMP, covering value added only in the 'material' sectors of production and excluding services, is the largest share.[2] This comes close to the market economy's definition of national income. However, the CPE's national accounting does not include so-called non-productive services (the value of services in the non-material sphere, such as research and development, banking, insurance, education, health, administration, and military). Of course, individual socialist countries' NMPs can deviate from the 'standard' *material product system* (MPS) in the inclusion of certain services in 'material' production. North Korea's income accounting includes the so-called 'transaction revenue' (alternatively, 'profits in turnover'), as does that of the former Soviet Union. The transaction revenue is equivalent to either excise taxes or some difference between retail price and wholesale price in a market economy.[3]

It is worth noting, therefore, that when we attempt to scale up NMP or socialist national income to estimate total gross national product (GNP), the CPE data must be assessed, and if necessary adjusted, for statistically significant problems of deviation from the 'standard' MPS, for the inclusion of certain services in 'material' production. But lack of data, and sometimes also lack of knowledge of production boundaries or valuation in the NMP practices of North Korea, have made it difficult to know whether or by how much to adjust its NMP or national income to an equivalent GNP or national income.

The second pitfall involves the interpretation of the self-serving and often falsified statistics put out by each country, both about itself and about the other. Particularly for North Korea, the reliability of official statistics released to the outside world is not certain. Problems of conceptual income accounting can occur because of the double counting of inputs, inadequate depreciation allowances, inclusion of wasted materials, and inadequate valuation of production. It is well known that in centrally planned economies prices are established administratively and satisfy neither the factor cost standard, which best reflects production potential, nor the welfare standard, which mirrors most accurately the contribution of goods and services to welfare.[4] It is not clear whether the administratively established prices are being used for valuation of all components of outputs or for either relatively high or low prices put on certain goods and services.

One important problem may be the inconsistent methods of imputation used to account for production that takes place outside the normal buyer–seller relationships and hence has no monetary value assigned to it. In most CPEs the prime examples are agricultural production consumed in kind, and the rental value of housing. It is not known if North Korea is always consistent in valuing these activities at prices at which equivalent goods or services are sold or produced. An equally important factor is the exchange rate that is needed to convert domestic currency income into an internationally convertible currency income, by which a one-to-one mapping of two countries' incomes is possible. But foreign exchange rates (say, the conversion rates of domestic currencies into US dollars) of both North and South Korea have been determined rather arbitrarily and unrealistically in accordance with international market conditions and the internal political and economic environments that the two countries face. South Korea used to fix its foreign exchange rates arbitrarily to help achieve equilibrium in external payments. Prices were set and changed independently of price movements in other countries until the mid-1970s. More recently, however, its foreign exchange rates are fully floating in line with supply and demand relationships in foreign currency markets. The official market rates are no longer lower than the equilibrium rates in the black market, if indeed the latter market still exists. The official exchange rates of North Korea, on the other hand, have remained almost unchanged ever since the country's establishment in the late 1940s. The response of North Korea to the US dollar devaluations of 1971 and 1973 and to the subsequent floating exchange rate system was obscure and unsynchronized.

As mentioned in Chapter 2, the main exchange rates in North Korea are the official exchange rate, the commercial (trade) exchange rate, and the non-commercial (tourist) exchange rate. Several other types of conversion coefficients may also be found, such as the exchange rates used in foreign currency shops (called 'Paradise Shops' in Pyongyang), those used for private remittances, and of course the black market rates.

Third, it must be pointed out that North Korea, like other command economies, is extremely biased towards the heavy industry and defence sectors, to the neglect of consumer goods.[5] Heavy industry production accounts for the major share of the income accounts, but it is of little true worth in terms of consumer welfare. The South Korean economy, on the other hand, has produced both a high gross national product (GNP) and a gross national pollution during the last three decades. The trade-off between income and pollution poses a serious problem in using real national income as a simple measure of the public's welfare. The same is true for North–South Korean comparison. There has also been a dangerous trend recently towards social schisms, towards economic and political polarization between urban workers and farmers, and between regions in South Korea.

Fourth, growth rates have often been used to compare economic performance between countries, but here too there are hidden hazards. The official growth rates of centrally planned economies are calculated with distorted prices and according to methods of index number construction that tend to yield varying degrees of upward bias. The degree of this bias is considered to be so substantial for some centrally planned economies, including North Korea, as to invite doubt about the validity of comparing the official NMP growth rate or production growth rates with the GNP growth rates of market-type economies.

Centrally planned economies rely on the double-deflation

method: the value of gross output and the value of material purchases and depreciation are deflated separately by the relevant price indexes. Gross value is typically overstated because new products generally tend to be priced higher, while the price indexes used to deflate them are downward-biased. Thus, constant-price output will have a higher arbitrary value, in general. The lack of independence of the central statistical offices in CPEs appears to be the source of such subordination of statistical integrity to political considerations. North Korea seems to be no exception.

To overcome such shortcomings, alternative measures have been attempted, using official data exclusively but as much as possible replicating commonly accepted standards of valuation and index number construction in market economies. The basic approach is to aggregate the official physical output series into sector and GNP indexes using weights constructed from official data.[6]

None the less, the implementation of this alternative approach is barely possible because of the serious lack of published data for North Korea and some other CPE countries. More importantly, the non-availability of published data on physical output as well as on monetary values is too great for any comparative study. There is a wealth of published data and statistics from South Korea, but the data from North Korea are extremely limited and are not more recent than the 1960s and 1970s. The country publishes only fragmentary information, with very little explanation of the methods used to compile the data.

North Korea began in the early 1950s to publish economic statistics, but largely in relative index numbers without specifying base-year quantities. The publication of such statistics declined from the mid-1960s, and since then there has been only the occasional release of fragmentary output statistics in terms of growth rates, but without any benchmark quantity figures, and they are mostly confined to industries and commodities that have met the plan's targets. This does not mean, however, that the statistics can simply be considered as unrealistic propaganda. They are the only official statistics known to the outside world, and they can yet provide some important information to be used for a comparison of the economic situations of the two states.

As a result of shortcomings in the official data from North Korea, considerable efforts have been devoted by Seoul and some Western sources, but the resulting aggregate estimates differ widely depending on the degree of credence given to official numbers and the use of divergent indexes and techniques. However, it is worth noting in passing that many experienced observers have rated North Korea's achievement quite highly. Harrison Salisbury, a very experienced observer of communist countries, visited the DPRK in 1972 and wrote of 'a tremendous technical and industrial achievement'; 'on a per capita basis it is the most intensively industrialized country in Asia, with the exception of Japan'.[7] The French agronomist René Dumont claimed in the late 1960s that: 'In agriculture and probably industry, too, North Korea leads the socialist block.' And Joan Robinson wrote after her visit in 1964 that 'All the economic miracles of the postwar world are put in the shade by these achievements.'[8] These observations generally refer to the period of the mid-1960s and early 1970s, when the DPRK economy seems to have outpaced that of the ROK.

It is generally admitted that the Soviet-type command economy in North Korea experienced rapid growth until the early 1970s. What has happened since then? The answer needs some in-depth analysis. For the moment it can be generally said that, once a command economy reaches a certain, threshold, level of development, it will come to face 'complexity'. Quite naturally, beyond that threshold it will begin to exhibit diminishing marginal returns to the inputs of effort typical in the centrally planned economy. This stage is followed by a drop in morale, and productivity suffers. Such phenomena have been noted in North Korea since the early 1970s, as well revealed, at least indirectly, by stepped-up '*chuche* idea' indoctrination and other morale-boosting drives that must have been pursued intensively by political leaders to motivate the people.

Nevertheless, there are many other more recent reports supporting the assertion that the DPRK economy is quite robust even in the 1980s. Jon Halliday has written:[9]

If official Pyongyang claims are sometimes exaggerated and often very vague, Western and especially Seoul downgradings are usually also inaccurate. A well-informed Soviet source said in 1984 that industrial production 'almost quadrupled' in the previous decade. There are undoubtedly major structural problems. Transportation and infrastructure lag badly behind. There are about 4380 kilometers compared to 6435 kilometers in South Korea (as of the end of 1990) of railway track, which carry almost all the country's goods. Very little seems to move by road. There has been a big drive to electrify the system and by the end of 1984, 88% electrification was claimed, although outside estimates do not confirm this.

John Merrill of the US Department of State also observed that:[10]

the North Koreans pay considerable attention to the agricultural sector, and it is an area in which they have had some success. The World Bank estimates that North Korean agricultural production exceeds the basic nutritional requirements by over a quarter, although this figure may have dropped in recent years. There have been reports of food shortages and cuts in rations, but this probably reflects other problems—such as distribution and bad weather—rather than serious deficiencies in the agricultural sector itself. In general, the agricultural sector has pulled its weight, generating some export earnings and maintaining reasonably good growth.

Admittedly, the North's achievement in agriculture is impressive, especially given its unfavourable climate and terrain. Land available for agriculture was to be expanded through massive projects scheduled for completion in 1990, both through the reclamation of 300 000 ha of land from the west coast, and terracing the uplands.

The fact that the economy has had some two decades or so of high economic growth means that it has become soundly established, which does not preclude a later slowdown, or even decline. The economic race between North and South will continue until the two Koreas eventually merge.

In this chapter the two economies will be assessed comparatively, first in terms of macroeconomic (GNP) aspects, and second with respect to real living standards. To begin with, we need either to find or to estimate quantitative data on North Korea's macroeconomic variables such as GVSP and GNP. The next sections will be devoted mostly to estimation problems and the practical approaches to be employed.

3.2   THE MACROECONOMIES

It seems to follow that anyone who seeks to make a statistical calculation of social income is confronted with a dilemma. The income he can calculate is not the true income he seeks; the income he seeks cannot be calculated. From this dilemma there is only one way out; it is of course the way that has to be taken in practice. He must take his objective magnitude, the social income *ex post*, and proceed to adjust it, in some way that seems plausible or reasonable, for those changes in capital values which look as if they have had the character of windfalls. This sort of estimation is normal statistical procedure, and on its own ground it is wholly justified. But it can only result in a statistical estimate; by its very nature, it is not the measurement of an economic quantity.

'Income' by John R. Hicks[11]

Methodological considerations concerning the reliability of available data reveal that the comparability of North Korea and South Korea has many pitfalls, as reviewed in the previous section. In comparing the macroeconomic aspects of the two sides, the foremost question is the reliability and consistency of available data and estimates of national incomes, as well as growth rates.

With regard to South Korea, most official publications can be taken as read, but there are some exceptions, of course. Economic data with political implications such as income distribution, inflation, credit rationings, shadow interest rates, use of foreign capital, military expenditures, etc., have always been under government control, particularly until the early 1980s. Therefore, some care may be needed in the use of these published time-series data. This could also hold true for such essential economic data as GNP or national income per capita, since these depend on price indexes and foreign exchange rates (when income is converted into a foreign currency), which can easily be manipulated for political and economic reasons. However, South Korea is an open country, and its economic performance is closely watched from outside by bodies such as the World Bank and major trading partners. Moreover, the country's methods of compiling and measuring national income and other statistics have conformed largely to the UN guidelines for the 'Measurement of National Income and the Construction of Social Accounts' and 'A System of National Accounts and Supporting Tables (SNA)'. Therefore, the extent of any falsification, omission, or error in statistical coverage is likely to remain minimal.

With regard to North Korea, on the other hand, the questions of credibility, quality, and feasibility are often raised with regard to various estimates of gross national product, not to mention other official data. From limited published data, a number of agencies in South Korea and in some Western countries have attempted estimates of North Korea's gross national product. Because of the serious methodological and practical problems involved in obtaining data on national product and growth rates for North Korea, these estimates vary widely, by a factor of more than 1.5. The principal causes of the differences appear to lie in two problems. The first is the differing concepts of national income and of the products on which estimates are based. The second is related to diverse values used in measuring national output, foreign exchange rates and price indexes, etc.

Before proceeding to the aggregate economic comparisons between North and South, it is worthwhile to review briefly the existing sources of estimates of North Korea's national income. This is then followed by a description of the particular methodology (and its rationality) used in this study for estimating North Korea's national income so as to make comparability 'possible with one yardstick'. A comparison of production of selected products between the two states will then be made.

3.2.1 Available Sources of North Korea's National Income Estimates

The Democratic People's Republic of Korea (DPRK) uses two macroeconomic indexes: the concept of gross social product (GSP), and national income (NI). As explained earlier, these macroeconomic concepts in the communist economy deviate from their counterparts in the market-type economy of the South. North Korea neither explains the methods of estimation of its macroeconomic variables, nor releases any details of the database used for the calculation of indexes. The only piecemeal bits of information known to the world are the growth rates of national income and of some arbitrarily selected commodities, but without base-quantity data, and some occasionally released per capita income figures.

Regarding the methodology of estimating GSP and NI, however, the *Economic Dictionary* published by the North Korean Academy of Social Sciences can be one source of information. In this official publication the gross social product is defined as 'material production aggregates being produced in such productive sectors of the society as [manufacturing] industry, agriculture, construction,

TABLE 3.1 Productive and non-productive sectors in North Korea

Productive Sector	Non-Productive Sector
Manufacturing industry	State administration
Agriculture	Public facilities management
Construction	Education, science, and art
Commodity transport	Health and welfare administration
Productive communication	Housing
Productive commerce	Passenger transport
Other (harvesting wild	Communication service for people
fruits, etc.)	Non-productive commerce

and commodity shipping and transportation'.[12] Accordingly, economic activities are classified according to whether they are productive or non-productive. The social wealth is considered to be created only by the productive sectors. The major division between productive and non-productive sectors is shown in Table 3.1. The GSP (or GVSP: gross output value of social production) adds the output values of all separately enumerated production units, as already explained briefly above. For example, suppose that a particular farm co-operative produces 10 units of wheat, of which 2 units are consumed by the co-operative itself. A wheat mill uses 8 units of wheat to produce 20 units of wheat flour, of which 5 units are consumed within the mill. Next, a bakery purchases 15 units of wheat flour to make 30 units of bread and it consumes 10 units of bread out of this 30 units. Then the gross social product earned in these productive processes is 43 units: $(10-2) + (20-5) + (30-10) = 43$. But in terms of the gross national product concepts used in market-type economies, this would come to 37 units: $10 + (20-8) + (30-15) = 37$. GNP accounts for all 'value added' in the production process, while GSP takes into account the *net* output values of all independent production units involved. In other words, the GNP concept subtracts the values of 'intermediate products' that are 'material inputs costs', and thus GNP is nothing but the sum of all 'value added', whether it is self-consumed or not. But the GSP (GVSP) includes the values of intermediate products for consecutive production processes, and thus their value is counted several times, while excluding only the values of self-consumed products.

North Korea's gross value of social product consists of three components in accordance with Marxian notation: $GVSP = C + V + S$, where in Marx's terminology C is constant capital, which represents productive equipment (factories, machinery, etc.), raw material and power; V is variable capital, which represents the wage bill; and S is the surplus value of labour, which represents 'net social income'. Note that here $C + V + S$ represents the flow of production per period of time; therefore, C is not the stock of capital invested, but the annual wear-and-tear and amortization of capital. In a capitalist society S constitutes the profits on capital invested and accruing to the factor income of the capitalists, but in North Korea (as in other communist countries) S is defined as the so-called 'surplus value', which is returned to society as a whole.

North Korea classifies 'net social income' into two categories: net income of business enterprises, and net income of central government. The net income of business enterprises is used for the expansion of production at enterprise and co-operatives level, and also for the compensation payments to workers and for social welfare. The state net income, which consists of 'profits in turnover' (transaction revenue), revenue of state enterprises, insurance fees, etc., is used for society as a whole.

Next, national income is defined in North Korea as a gross value of social product minus capital depreciation and intermediate costs. This definition is certainly based on 'the theory of labour value' in Marxian doctrine, where production values come only out of labour inputs, not from other inputs such as capital and land. In North Korea, capital is regarded as nothing more than the outcome of the accumulation of past labour inputs.

Major differences between the GNP and GSP (GVSP) concepts may be summarized as follows. First, GNP includes both productive and non-productive sectors in its measurement, whereas in GSP the non-productive sector is excluded. Second, GNP is the sum of 'value added' (total value of outputs minus intermediate costs), while GSP includes intermediate (material) inputs except for goods consumed within the same enterprise. As with GSP (GVSP) measurements, North Korea's national income does not include non-productive sectors, but does include the share of 'profits in turnover' (transaction revenue). The concept of national income in a market economy is actually expressed in terms of 'factor income' at 'factor prices'. But in North Korea national income is measured in terms of actual 'transaction (market) prices'.

In order to make economic comparison possible, therefore, overall indicators of North Korea must be somehow adjusted to conform as much as possible to those measures used in capitalist economies. There are a number of problems involved in doing this, particularly if accounting for so-called non-productive factor incomes and subsidies as well as transaction revenue (turnover profits). (North Korea regards transaction revenue a part of net social income—that is, profits of the state, not 'sales taxes'— mainly to emphasize that there is no tax in the DPRK.) Not-withstanding, a number of individuals and agencies in South Korea[13] and in some Western countries[14] have attempted to estimate North Korea's GNP or per capita income. These estimates vary widely, however, partly because of differing concepts and methods used,[15] and partly because of different values of data in use.

Table 3.2 presents estimates of North Korea's gross national product prepared by some selective sources.[16] However, some of those original estimates either were in North Korean currency or were not in complete form in each series, and so I have carried out some arbitrary extrapolation and conversion into US dollars (for details, see table footnotes).

Putting aside North Korea's GSP (GVSP) estimates in column (1) of the table, GNP figures vary dramatically even if they are compared among cohort groups that used a unique conversion factor. For example, the GNP estimates of the National Unification Board (NUB) and Joseph Chung are both based on the North Korean trade exchange rate, while IEAS, FSO, and the US CIA's figures are based on the official exchange rate. Lee's estimate, originally in North Korean currency, is converted here into US dollars by using his recommended purchasing power parity (PPP) rate of US$1 = W1.66. In the first cohort group, Chung's estimates are slightly lower than those of the NUB until 1980, but thereafter his figures run ahead of the NUB estimates. In the second cohort group, estimates range from $1000 million (CIA) to $6253 million (IEAS) for 1970, and for 1979 estimates range from a high of $31544 million (FSO) to a low of $17336 million (CIA). Taking as an example the estimates for 1984, the disparity ratio is almost 1.56 between the highest value (CIA) and the lowest one (NUB), disregarding the exchange rates used.

Such incongruous estimates may be attributed to both personal judgement, based on scanty documented quantitative data, and to the estimation methods used. Since each estimate appears to have a wide variance from the true value, the use of the mean value of the amalgamated estimates as a proxy for North Korea's true GNP would be seriously incommensurate and is not recommended. This study will take the estimated North Korean GVSP (GSP) as a departure point from which to estimate an approximately comparable GNP.

3.2.2   North Korea's GNP Estimation using GVSP

Generally, five alternative approaches[17] may be considered with regard to an estimation of GNP in centrally planned economies. The first approach entails building a more or less complete set of national accounts from disaggregated data, and then computing GNP as the sum of value added in the production sectors—industry, agriculture, and the like. This approach needs production data aggregated according to some sort of factor cost valuation.

The second approach is scaling up from net material product (NMP) to GNP on the basis of an average relationship found to hold for either other communist countries or Western countries. Differences in economic systems as well as in qualities and levels of development must be taken into consideration.

The third approach relates to scaling up from NMP by adding net value added in the non-material sectors plus depreciation, and otherwise adjusting as necessary to make the GNP estimates for centrally planned economies comparable with those for market economies. This is actually the best practical method, so long as clearly defined sectoral production data including those from the non-productive service sector are available from official sources. Otherwise, the method is subject to estimation errors, and is also time-consuming because of the need for data reconstruction and adjustments.

The fourth method is to derive GNP as the sum of the end-use values of each economic unit—consumption, investment, government, and net exports. This is an expenditure approach which is widely used in estimating income in market economies.

A fifth method is called the 'physical indicators approach'. This involves estimating GNP directly in dollars via currency conversion

TABLE 3.2 Differences in Estimates of North Korea's GVSP and GNP (US$ million)

	(1) North's GVSP[a]	(2) NUB[b]	(3) US CIA[c]	(4) ACDA[d]	(5) SIPRI[e]	(6) IISS[f]	(7) IEAS[g]	(8) FSO[h]	(9) Chung[i]	(10) Lee[j]
1946	677									325
1949	1 482									681
1953	1 131									476
1960	5 568	1 520	4 800				2 708	3 587		2 229
1961	6 301	1 810					2 939			2 892
1962	6 999	2 020					3 181			2 837
1963	7 659	2 150		2 300			3 461			3 265
1964	8 425	2 300		2 500			3 894			3 566
1965	8 734	2 340	7 600	2 500			4 247	5 768		3 807
1966	9 241	2 410		2 900			4 427			4 060
1967	9 777	2 600		3 000			4 985			
1968	10 344	2 980		3 500			5 534			
1969	10 944	3 120		4 000			5 673			
1970	14 334	3 980	10 000	4 500		8 510	6 253	9 132	4 010	
1971	17 978	4 090		4 900			7 459		4 090	
1972	20 955	4 620		5 300			8 273		4 620	
1973	24 817	6 270		—					6 050	
1974	33 631	7 290		8 595					7 040	
1975	40 357	9 350	16 000	9 975		7 980		19 769	9 260	
1976	44 796	9 680	15 550	10 471				21 282	9 500	
1977	43 004	10 640	15 550	10 781				22 796	10 180	
1978	51 938	13 320	16 685	—				26 003	13 000	
1979	66 128	16 040	17 336	—				31 544	14 770	
1980	65 688	13 500	19 500	18 050	14 330	13 370			13 530	
1981	67 331	13 560	21 400	19 810	14 580	16 310			14 390	
1982	66 978	13 600	22 200	20 500	15 310	15 610			15 930	
1983	71 466	14 770	22 400	20 720	16 000	17 570			16 030	
1984	73 395	14 720	23 000		19 580	18 740			17 410	
1985	71 854	15 140			20 160	19 240			17 950	
1986	76 960	17 350			20 610				18 990	
1987	86 265	19 400							20 450	
1988	88 853	20 600								
1989	88 688	26 494								
1990	90 727	27 100								

[a] For 1946–65, North Korea's announced GVSP figures in current prices were converted into US dollars by the official exchange rate of $1 = W1.20. For 1966–81, GVSP data were estimated by using North Korea's growth rates estimates by US CIA (5.8% in 1966–9; 31.0% in 1970; 16.0% in 1971–2; 19.0% in 1973; 17.2% in 1974; 20.0% in 1975; 11.0% in 1976; −4.0% in 1977; 17.0% in 1978; 15.0% in 1979; 1.7% in 1980; 2.5% in 1981); and these figures were converted into US dollars by official exchange rates given in Table 2.10. For 1982–3, GVSPs were estimated, assuming annual average growth rate of 12.2%, which was the official average growth rate of industrial production during the second Seven-Year Plan (1978–84). For 1984 and 1985, a growth rate of 2.7% was used. For 1986 and 1987, growth rates of 2.1% and 3.3% were applied respectively. For 1988 and 1989, a growth rate of 3.0% was used. For 1990, a growth rate of 2.3% is applied.

Table notes continued

These estimated GVSP (or GSP) figures differ from the other estimates which are all in terms of GNP concepts, and in converting GVSP (GSP) into dollars, official exchange rates (given in Table 2.10) were used, while in GNP estimations mostly trade or non-trade rates were used with exceptions of IEAS, FSO, and Lee's estimations.

[b] GNP estimated by South Korea's National Unification Board (NUB). These figures were obtained by dividing North Korea's official budget data by a factor of 0.60 and were converted into US dollars by the trade exchange rate given below:

1946– 72	1973– 4	1975	1976– 77	1978	1979	1980– 1	1982	1983	1984	1985	1986	1987	1988
2.57	2.37	2.05	2.15	1.86	1.79	1.70	2.12	2.18	2.36	2.43	2.23	2.14	2.10

[c] US CIA estimates of North Korea's GNP were based on 1984 constant US dollars. The CIA is known to have used North Korea's official exchange rate as the conversion factor. However, the CIA *Handbook of Economic Statistics* (1984, 1985) reported average exchange rates of North Korea as follows (non-commercial rate to 1977; 1978 is a special rate for March 1978 as reported; data for 1981–2 are estimated):

1971	1973	1975	1976	1977	1978	1979	1980	1981	1982	1983
2.36	2.06	2.02	2.02	2.02	1.86	1.79	1.77	1.85	2.00	2.13

[d] US Arms Control and Disarmanent Agency, *World Military Expenditures and Arms Transfers, 1967–76, 1984, 1985* (US Government Printing Office). These figures are in 1970 US dollar prices for 1963–77, and in 1982 US dollars for 1980–3.

[e] Stockholm International Peace Research Institute, *Yearbook 1987: World Armament and Disarmament.* Estimates of the GNP were made by using the ratio of defence expense to GNP in 1980 constant exchange rate.

[f] International Institute of Strategic Studies, *The Military Balance, 1987–88* (Washington, DC, 1989). The estimates were made on the 1984 official exchange rate, i.e. $1 = W0.94.

[g] Institute for East Asian Studies in Seoul estimated North Korea's GNP in accordance with SNA method, 1960–72, and in terms of North Korean currency. These estimates were converted into US dollar terms by official exchange rates in Table 2.10.

[h] Federal Statistical Offices (West Germany), *Statistik Des Auslaendes: Korea, Demokratische Volksrepublik* (1986), 63, 64. These figures are in official exchange rates.

[i] See Joseph Chung, 'A Study of North Korean Economy', *Korea and International Politics*, 5:2 (Kyongnam University, 1989). The estimates of North Korea's GNP are in terms of trade exchange rates.

[j] See Pong S. Lee, 'An Estimate of North Korea's National Income', *Asian Survey*, 12:6 (University of California, 1972), 518–26; see Lee's table 2 on p. 522. Originally Lee estimated North Korea's national income in 1966 North Korean currency (*won*) for the period 1946–67. His estimate is converted into US dollars by using his recommended rate of purchasing-power parity at $1 = W1.66. However, note that Lee's estimate is national income (NI), not GNP. It can be roughly approximated that NI is around 90% of GNP, taking account of 3.7% of gross production as depreciation cost and 5.5 to 7.4% as the factor income ratio of nonproductive sectors.

(for example, using the prevailing exchange rate or some type of purchasing-power parity). The essence of this approach is the determination of a regression relationship between a set of physical indicators (such as consumption and the stock of certain assets) of development and the per-capita dollar gross domestic product (GDP) of arbitrarily chosen sample countries. The statistical relationship obtained is then used to estimate the per capita dollar GDP for the country where neither GDP nor a meaningful dollar exchange rate could be found. This approach sounds well grounded

TABLE 3.3 North Korean GVSP, 1946–1965 (W million)

	GVSP		GVSP		GVSP
1946	812.5	1960	6682.0	1964	10 110.0
1949	1779.4	1961	7560.9	1965	10 481.2
1953	1357.7	1962	8398.3		
1956	2856.2	1963	9191.0		

Source: *North Korea's Statistics, 1946–85* (Seoul: National Unification Board, 1986), 124.

and practical, but it too encounters the problems of structural and development-level differences between the target country and reference countries.

None of these methods could be applied to North Korea, however, because no comprehensive, consistent, and up-to-date statistical information on that country's NMP or physical aggregate data could be obtained or reconstructed for direct use.

This study shall turn hereafter to a relatively simple and practical method of obtaining an estimate of approximately comparable reliability from GVSP (GPS) estimates like one shown in Table 3.2. The numbers in the first column of the table are based partly on the NUB's estimates of North Korea's GVSP (1946–65),[18] which used both the GVSP growth index (1946–65) and production data given in *Chosen Chungang Nyungam* (North Korea's Central Yearbook) and DPRK's *Statistics 1946–1960* (see Table 3.6). The estimates were extrapolated to 1989 by using the growth rates of the North Korean economy. (See note *a* of Table 3.2 for the sources of data.)

The key problem here lies in the use of the growth rates of the industrial sector for the extrapolation of North Korea's GVSPs, particularly for 1982–3. There is some possibility that the growth of industrial production alone may result in overestimates of increases in GVSP; for if the agricultural sector lags too far behind, it can severely retard overall growth as long as its contribution to the country's GVSP is very large. But the share of the agricultural sector in North Korea's GVSP dropped from 41.6% in 1953 to 26.6% in 1956 to 23.6% in 1960, and to less than 19% since 1964. The composition of North Korea's GSP in selected years is given in Table 3.7.

TABLE 3.4 Pong S. Lee's estimates of North Korean GVSP, 1946–1967 (1966 W million)

	Est. GVSP		Est. GVSP		Est. GVSP
1946	920	1960	7 300	1965	12 100
1949	2 000	1961	8 600	1966	13 000
1953	1 500	1962	9 100	1967	n.a.
1956	3 200	1963	10 000		
1959	6 700	1964	11 000		

Source: Pong S. Lee, 'An Estimate of North Korea's National Income', *Asian Survey*, 12:6 (University of California Press, 1972), 522.

TABLE 3.5 Pong S. Lee's estimates of North Korean national income, 1946–1967 (1966 W million)

	NI		NI		NI
1946	540	1960	3700	1964	5420
1949	1130	1961	4800	1965	5920
1953	790	1962	4710	1966	6320
1956	1730	1963	5030	1967	6740

Source: Lee, 'Estimate', 519.

The average annual growth rates of North Korea's gross agricultural product were 10.0% for 1954–60 and 6.3% for 1961–70.[19] The growth rates for the agricultural sector were estimated to be relatively low for the period 1971–8, averaging 1.6% per year.[20] The agricultural sector has received major focus on modernization since the late 1970s, however, and is believed to have grown at an average rate of more than 6.2% during 1982–3.[21] Therefore, for 1982–3, GVSPs shown in column (1) of Table 3.2 were possibly overestimated by a negligible factor of 1.16% at most. The *Asia Yearbook* (1984) gave average growth over the years 1978–82 at 9.7% for North Korea—the highest in Asia, except for Brunei.[22]

The second focal point concerns an appropriate rate of exchange. Since North Korea operates multiple exchange rates, which can diverge by a ratio of up to 3 : 1, the choice of a conversion ratio will dramatically affect the estimates, as demonstrated in Table

TABLE 3.6 Major official statistics of North Korea's economy, 1946–1964

	1946	1947	1948	1949	1950	1951	1952	1953
GSP (%)	100	225.6	260	219	—	—	—	163
Industry	23.2	—	—	35.6	—	—	—	30.7
Agriculture	59.1	—	—	40.6	—	—	—	41.6
Transport and communication	1.6	—	—	2.9	—	—	—	3.7
Basic construction	—	—	—	7.2	—	—	—	14.9
Commerce and marketing	12.0	—	—	9.4	—	—	—	6.0
Others	4.1	—	—	4.3	—	—	—	3.1
National income	100	—	—	209	—	—	—	145
Labour productivity	100	—	—	252	—	—	—	197
Government revenue	100	—	—	13-fold	—	—	—	32-fold
Government expenditure	100	—	—	15-fold	—	—	—	37-fold
People's Economy (growth rate)	100	—	—	35-fold	—	—	—	100-fold
Culture and social (growth rate)	100	—	—	13-fold	—	—	—	19-fold
Population ('000)[a]	9257	—	—	9622	—	—	—	8491

[a] Based on the growth of labour productivity in industry.

Source: *Chosen Chungang Nyungam* (North Korea's Central Yearbook) (1961, 1964, 1965).

3.2. The fairest measure is probably purchasing price parity (PPP), but unfortunately, it is virtually impossible to estimate North Korea's PPP because of insufficient up-to-date information and data.[23] Therefore, this study chooses to take the official rather than the trade (commercial) or non-trade (tourist) rate, or other surrogate rate.

Third, in order to get some basis for comparison between different estimates, additional estimates such as GVSP(II) and GVSP (III) are added in Table 3.8. GVSP(II) is partially borrowed from Pong S. Lee's estimate for 1946–67, and is then extrapolated up to 1987 using the same growth rates as used in the estimation of GVSP(I). GVSP(III) is methodologically somewhat different. In the first place, on the basis of several piecemeal and sporadic

1954	1955	1956	1957	1958	1959	1960	1961	1962	1963	1964
—	—	355	—	—	735	797	941	10-fold	11-fold	12-fold
—	—	40.1	—	—	—	57.1	—	61.0	60.6	62.2
—	—	26.6	—	—	—	23.6	—	21.3	21.5	19.3
—	—	4.0	—	—	—	2.2	—	2.8	2.8	2.8
—	—	12.3	—	—	—	8.7	—	9.1	9.2	9.8
—	—	10.8	—	—	—	6.0	—	3.6	3.8	3.8
—	—	6.2	—	—	—	2.4	—	2.2	2.1	2.0
—	—	319	417	594	—	683	—	869	928	10-fold
—	—	386	477	506	—	539	—	—	631	709
—	—	61-fold	—	—	—	123-fold	—	—	192-fold	214-fold
—	—	71-fold	—	—	—	146-fold	—	—	224-fold	253-fold
—	—	298-fold	—	—	—	570-fold	—	—	938-fold	996-fold
—	—	41-fold	—	—	—	162-fold	—	—	217-fold	234-fold
—	—	9359	—	—	—	10 789	—	—	11 717	—

TABLE 3.7 Composition of North Korean GSP, selected years, 1953–1983 (%)

Sector	1953	1956	1963	1970	1983[a]
Industry[b]	30.7	40.1	62.3	64.2	66.0
Agriculture[b]	41.6	26.6	19.3	18.3	10.0
Transport and communication	3.7	4.0	2.8	—	—
Basic construction	14.9	12.3	9.8	—	—
Commerce and other material supplies	6.0	10.8	3.8	7.5	—
Others	3.1	6.2	2.0	—	—

[a] Estimated.
[b] Note that the industrial sector in GSP (GVSP) may tend to be overestimated and the agricultural sector underestimated from the viewpoint of GNP, because of multiple counting in industry and the fact that there is no turnover tax (transaction revenue) in the agriculture sector.

Sources: *Chosen Chungang Nyungam, 1965* (Pyongyang, 1965), 477; *Rodung Sinmun* (27 January 1984).

1975	38 742.64	45 419.61	34 659.40	40 356.91	47 312.09	36 103.54
1976	43 004.00	50 416.76	36 865.62	44 796.06	52 517.45	38 401.68
1977	41 284.05	48 399.13	39 212.38	43 004.21	50 415.76	40 846.22
1978	48 302.34	56 626.98	41 708.44	51 938.00	60 889.22	44 847.78
1979	55 547.69	65 121.03	67 378.92	66 128.20	77 525.03	80 213.00
1980	56 492.01	66 228.09	71 776.98	65 688.38	77 009.40	83 461.60
1981	57 904.31	67 883.79	78 102.18	67 330.59	78 934.63	90 816.48
1982	64 968.63	76 165.61	83 260.04	66 977.96	78 521.24	85 835.10
1983	72 894.81	85 457.82	83 063.36	71 465.50	83 782.17	81 434.66
1984	74 862.97	87 765.18	85 160.04	73 395.06	86 044.29	83 490.23
1985	76 884.27	90 134.84	88 612.36	71 854.45	84 238.16	82 815.28
1986	78 498.84	92 027.67	90 622.50	76 959.64	90 223.20	88 845.58
1987	81 089.30	95 064.58	93 648.78	86 265.21	101 132.53	99 626.36
1988	83 521.98	97 916.52	96 527.88	88 853.17	104 166.51	102 689.23
1989	86 027.64	100 854.01	100 563.10	88 688.26	103 973.20	103 673.29
1990	88 006.28	103 173.65	103 173.65	90 728.09	106 364.37	106 057.77

Note: GVSP (I) is based on the NUB's estimate of North Korea's GVSP (1946–65) as given in n. 19 of this chapter, while GVSP (II) is borrowed from Pong S. Lee's estimate (1946–66). These two estimates were extrapolated up to 1989 by using North Korea's growth rates as given in n. *a* in Table 3.2. Then they were converted into US dollars by using official exchange rates given in Table 2.11.

TABLE 3.8 Estimates of North Korea's GVSP, 1946–1990

| | GVSP (I) | GVSP (II) (millions of North Korean won) | GVSP (III) | GVSP (I) | GVSP (II) (millions of US dollars) | GVSP (III) |
	X_1	X_2	X_3	Y_1	Y_2	Y_3
1946	812.50	920.00	1 193.04	677.08	766.67	994.20
1949	1 779.40	2 000.00	2 540.20	1 482.83	1 666.67	2 116.83
1953	1 357.70	1 500.00	1 711.14	1 131.42	1 250.00	1 425.96
1956	2 856.20	3 200.00	5 486.04	2 380.16	2 666.67	4 571.70
1960	6 682.00	7 300.00	7 809.50	5 568.33	6 083.33	6 507.91
1961	7 560.90	8 600.00	8 613.16	6 300.75	7 166.67	7 177.63
1962	8 398.30	9 100.00	9 499.36	6 998.58	7 583.33	7 916.13
1963	9 191.00	10 000.00	10 178.02	7 659.16	8 333.33	8 481.68
1964	10 110.00	11 000.00	11 226.36	8 425.00	9 166.67	9 355.30
1965	10 481.20	12 100.00	11 807.18	8 734.33	10 083.33	9 839.31
1966	11 088.89	13 000.00	12 433.96	9 240.74	10 833.33	10 361.63
1967	11 732.04	13 754.00	14 821.46	9 776.70	11 461.67	12 351.67
1968	12 412.50	14 561.73	15 641.76	10 343.75	12 134.77	13 034.80
1969	13 132.42	15 395.73	16 208.68	10 943.68	12 829.77	13 507.23
1970	17 203.48	20 168.40	16 829.60	14 336.23	16 807.00	14 024.66
1971	19 956.03	23 395.35	18 634.66	17 978.41	21 076.89	16 787.98
1972	23 149.00	27 138.60	20 412.84	20 954.95	24 449.18	18 389.94
1973	27 547.31	32 294.94	22 481.24	24 817.40	29 094.54	20 253.36
1974	32 285.45	37 849.67	31 474.22	33 630.68	39 426.73	32 785.64

TABLE 3.9 North Korea's per capita income, 1946–1982 (official data)

	Per capita income (*won*)	Sources
1946	64.44[a]	W580 in 1967 is 9.0-fold of income of 1946
1949	132.00[a]	W580 in 1967 is 4.4-fold of income of 1946
1962	416.66[b]	W500 in 1966 is 1.2-fold of income of 1962
1966	500.00[b]	Report of Supreme People's Convention (16 Dec. 1967)
1967	580.00[b]	Report of Supreme People's Convention (16 Dec. 1967)
1970	605.73[c]	Income in 1970 is 9.4-fold of income of 1946
1974[e]	1020.69[d]	Income in 1974 is 1.7-fold of income of 1970
1974	1000.00[d]	Industrial Workers' Meeting (4 March 1975)
1979	1920.00	Kim Il-Sung's New Year Message (1 Jan. 1980)
1982	2200.00	Kim Woo-Jung's News Conference with Japanese Reporters (13 Sept. 1983)

[a] *Chosen Chungang Nyungam* (1970), 276.
[b] *Chosen Chungang Nyungam* (1968), 2.
[c] *Chosen Chungang Nyungam* (1974), 242.
[d] *Chosen Chungang Nyungam* (1976), 30, 371.
[e] In 1974 per capita income estimated by North Korea's official growth rate deviates from data released at Industrial Workers' Meeting held on 4 March 1975. In Table 3.8 I used the estimated figure of 1020.69 as a basis for estimating each subsequent year's income until 1979, when new data were released for the year by Kim Il-Sung.

announcements made concerning per capita income and growth of income in addresses such as those made by Kim Il-Sung and other high-ranking officials in connection with special occasions, data for nine benchmark years were established (see Table 3.9). Next, using these hierarchical benchmark data and average annual growth rates of per capita real income estimated by the US CIA and myself (see notes to Table 3.10), yearly data series of per capita income are lexicographically built up. Then, multiplying per capita income series by annual population estimates produces yearly national income, which is given in the last column of Table 3.10. Lastly, assuming that North Korea's GVSP is approximately twice the national income, GVSP(III) is derived as shown in Table 3.8.

Surprisingly, there do not seem to be many variations among the three estimates of North Korea's GVSPs. But to choose the most probable of the three alternatives, the principal components

TABLE 3.10 Estimates of North Korea's national income used to calculate GVSP (III) in Table 3.8

	Per capita income (W)	Average annual growth rate of per capita real income (%)[a]	Population[b]	National income[c] (W million)
1946	64.44[d]		9 257 000	596.52
1949	132.00[d]		9 622 000	1 270.10
1953	100.76		8 491 000	855.67
1956	293.09	23.4	9 359 000	2 743.02
1960	361.92	15.8	10 789 000	3 904.75
1961	388.33	6.8	11 090 013	4 306.58
1962	416.66[d]		11 399 424	4 749.68
1963	434.31	6.8	11 717 468	5 089.01
1964	466.00	6.8	12 044 385	5 612.68
1965	487.90	4.7	12 100 000	5 903.59
1966	500.00[d]		12 433 960	6 216.98
1967	580.00[d]		12 777 137	7 410.73
1968	595.66	2.7	13 129 786	7 820.88
1969	600.67	0.8	13 492 168	8 104.34
1970	605.73[d]		13 892 000	8 414.80
1971	649.94	7.3	14 258 748	9 267.33
1972	697.39	7.3	14 635 179	10 206.42
1973	748.30	7.3	15 021 548	11 240.62
1974	1020.69[d]		15 418 117	15 737.11
1975	1093.15	7.1	15 853 000	17 329.70
1976	1133.60	3.7	16 260 422	18 432.81
1977	1175.55	3.7	16 678 314	19 606.19
1978	1219.05	3.7	17 106 947	20 854.22
1979	1920.00[d]		17 546 596	33 689.46
1980	1991.04	3.7	18 025 000	35 888.49
1981	2114.48	6.2	18 468 415	39 051.09
1982	2200.00[d]		18 922 738	41 630.02
1983	2164.80	−1.6	19 185 000	41 531.68
1984	2169.13	0.2[e]	19 630 000	42 580.02
1985	2173.47	0.2[e]	20 385 000	44 306.18
1986	2169.13	−0.2[e]	20 866 085	45 261.25
1987	2190.82	1.0[e]	21 373 000	46 824.39
1988	2206.15	0.7[e]	21 877 000	48 263.94
1989	2232.63	1.2[e]	22 521 223	50 281.55
1990	2219.60	−0.1[e]	23 174 434	51 438.02

[a] Derived partially from US CIA, *Handbook of Economic Statistics* (1986), 40. Some of the growth rates were estimated by the author, based on North Korea's official per capita income data (as shown in Table 3.9) and extrapolated data.

[b] Population data for 1946–63 were from North Korea's official data (*Chosen Chungang Nyungam* (1964), 316); data for 1965, 1970, 1975, 1980, 1983–5, 1987, 1988, and 1990 were borrowed from UN, *World Population Prospects: Estimates and Projections As Assessed in 1984* (1991), 50–1. Data for 1989 came from CIA, *The World Factbook* (1989), 163. Data for other years were estimated on the basis of average annual growth rate of population in North Korea. (See UN, *World Population Prospects* (1991), 56–7, for annual rate of growth.)

[c] Note that the communist definition of national income excludes most services; it equals GVSP minus all material costs including depreciation of capital.

[d] Derived from Table 3.9.

[e] Growth rates of per capita real income for 1984–9 were estimated by the author.

are calculated from these three series.[24] Before calculating the principal components, simple correlation coefficients between every pair of series are estimated as follows:

	X_1	X_2	X_3
X_1	1.0		
X_2	0.9999	1.0	
X_3	0.9762	0.9756	1.0

where $X_1 = $ GVSP(I), $X_2 = $ GVSP(II), and $X_3 = $ GVSP(III), all expressed in millions of North Korean *won*. Similarly,

	Y_1	Y_2	Y_3
Y_1	1.0		
Y_2	0.9205	1.0	
Y_3	0.9770	0.9028	1.0

where $Y_i = $ GVSP(i), all expressed in terms of millions of US dollars which were converted from North Korean *won* using the official exchange rate. As can be seen from the simple correlation matrix, X_1 is correlated more with X_2 than either X_1 is with X_3 or X_2 is with X_3. But in US dollar terms, the correlation between Y_1 and Y_3 is significantly larger than the relationship between either Y_1 and Y_2 or Y_2 and Y_3. However, these three series are very interdependent, and it is almost impossible to determine which series to choose simply by looking at the correlation coefficients.

As an alternative approach to determining the search criterion, each of these three series is regressed on the remaining two, and the three multiple determination coefficients are obtained for each X_i and Y_i series, as follows:

Dependent variable	Multiple determination coefficient	Dependent variable	Multiple determination coefficient
X_1	0.9999	Y_1	0.9812
X_2	0.9987	Y_2	0.9206
X_3	0.9766	Y_3	0.9770

Now we have a somewhat better picture, in that the X_1 series (similarly, the Y_1 series) is the most correlated with the other two series. Particularly in the case of the X_i series, X_1 is positively and significantly correlated with both X_2 and X_3, but X_2 is positive with

X_1 but negative and insignificant with X_3 in their estimated co-efficients. Also, X_3 is positively related with X_1, but negative and insignificant with X_2. The results for Y_i series were almost similar in terms of t-statistics of estimated coefficients.[25] Thus, the X_1 series (and Y_1 series) appears to be the prime candidate for choice. Now, three principal components are calculated for X_i series as well as for Y_i series, and the absolute sizes of variances and accumulated percentages of variance explanation are obtained for each of the principal components as follows:

	Absolute size	*Accumulated %*
X_1	2.98380	0.99460
X_2	1.01614	0.99998
X_3	0.00007	1.00000
Y_1	2.93208	0.97736
Y_2	0.05699	0.99636
Y_3	0.01093	1.00000

Then simple correlation coefficients between each of the GVSP series (that is, X_i and Y_i series) and the first principal components are again calculated as follows:[26]

X_1	0.9974
X_2	0.0028
X_3	0.0000
Y_1	0.9881
Y_2	0.0378
Y_3	0.0047

Considering the above correlation coefficients, the X_1 series (also the Y_1 series) is definitely considered the best approximation of North Korea's GVSP series, which, after reducing them to GNP content, can be compared with South Korea's GNP.

The next important task is to derive North Korea's GNP from GVSP(I), which we choose as the best possible approximation for North Korea's gross value of social production. We know that the difference between the socialist GVSP and the capitalist GNP lies mostly in the fact that GVSP(III) is based on the estimate of North Korea's national income in Table 3.10. Assuming that national income is approximately half of North Korea's GVSP, GVSP(III) is obtained by multiplying the national income estimate by a factor of 2. Another conversion factor is North Korea's official exchange rate.

GVSP includes the values of intermediate products, but excludes so-called non-productive services such as administration, education, health care, passenger transport, defence, and housing services. Estimating the value of these services is highly problematic. For example, it is not easy to value housing services in the socialist economy, because first of all a decision must be made as to what value should be put on each type of housing, both public and private (if any).

The usual procedure for estimating the contribution of non-material sectors is to begin with published figures on employment, average wages, and social security to derive the wage bill, then to add operating surplus in the form of profits and other types of accumulation as reported or as can be best estimated. But again, such a method cannot be applied to North Korea owing to lack of data and also the non-existence of information on material and non-material production boundaries within its loosely released data-set.

A practical method is therefore employed. First, note that it has already been assumed, in the calculation of GVSP(III) in Tables 3.8 and 3.10, that the NI/GVSP ratio in North Korea is about 0.5. Next, cross-country examination shows that the NI/GNP ratio falls within the range between 0.75–0.90 depending upon whether NI is measured at factor cost or market price. Taking South Korea as reference, the NI/GNP ratio was 0.78 at factor cost and 0.895 at market price in 1979. An examination of this ratio in both communist and capitalist developing economies shows an average of 0.8 over an extended period of time (1960–80). This, in turn, confirms my approximation that the GNP/GVSP ratio appears to be approximately 0.625 for North Korea.

Alternatively, observing the ratio of material and depreciation costs to GNP in South Korea gives us some numbers centred around 23.3% both in the late 1970s and for most of the 1980s. If this factor is used, about 37.3% of North Korea's GVSP is considered to be attributable to those proportions of double-counting, net of the service sector, that are left unaccounted. Therefore, North Korea's GNP in terms of market-type economic definition is equivalent to (1–0.373) GVSP; that is its GNP/GVSP ratio works out to about 0.63.

Now, the GNP derivation using North Korea's GVSP(I) as selected above becomes straightforward and the result is shown in Table 3.11(*a*), where dollar GNPs of North Korea are presented for the cases converted both by official exchange rate and by trade exchange rate, although I believe that it is more rational to use the official exchange rate in calculating dollar GNP because the share of the trade sector in the economy is relatively small (see Chapter 5). So in the following discussion I shall use North Korea's GNP estimate based on the official exchange rate in comparing with South Korea's GNP shown in Table 3.11(*b*). This official exchange rate GNP estimate conforms very well with North Korea's official GNP announcement as well as with other sources to be discussed in Section 3.3.

3.2.3   GNPs of North and South Korea

We have estimated North Korea's GNP (Table 3.11(*a*)) so as to allow comparison with South Korea's GNP (Table 3.11(*b*)). To begin with, however, it must be noted again that the estimation of per capita GNP in US dollar terms depends critically on the *won*–dollar exchange rate and on population estimates, not to mention the methods used in approximating the macro variables. For example, North Korea's GNP and its per capita GNP in US dollars are both less than half of the estimated figures based on the official exchange rate shown in Table 3.11(*a*) when we use North Korea's trade exchange rate.

Even if data problems such as omission and errors exist in both North Korea and South Korea, and given the dearth of statistics from North Korea, we now have a relatively reliable set of macro data for both states.[27] Cross-checking with other studies has shown that estimations in this study have an inner consistency with available data. As an example, South Korea claims that it overtook the North in per capita GNP in 1969.[28] This assertion would be nearly true if the trade exchange rate (as usually adopted by South Koreans) were used as a conversion factor, that is, if North Korea's per capita GNP was around $230 in 1969. Meanwhile, the US CIA claims that South Korea began to surpass North Korea in aggregate GNP around 1976. Surprisingly, this claim conforms well with our estimation based on the official exchange rate (see Tables 3.11(*a*) and 3.11(*b*)). In absolute magnitude of dollar GNP,

TABLE 3.11(a) North Korea's GNP and per capita GNP, 1946–1990

	GNP[a] (W million)	GNP[b] (US$ million)	GNP[c] (US$ million)	Per capita GNP[b] ($)	Per capita GNP[c] ($)	Growth rate[d] (%)
1946	511.9	426.6	232.7	46	25	—
1949	1 121.0	934.2	509.5	97	53	—
1953	855.4	712.8	388.8	84	46	—
1956	1 610.4	1 499.5	732.0	160	78	31.8
1960	4 209.7	3 508.0	1 913.5	325	177	7.4
1961	4 763.4	3 969.5	2 165.2	357	195	13.1
1962	5 290.9	4 409.1	2 404.9	386	211	11.0
1963	5 790.3	4 825.3	2 632.0	411	225	9.4
1964	6 369.3	5 307.8	2 895.1	440	240	9.9
1965	6 603.2	5 502.6	3 001.5	454	248	3.6
1966	6 986.0	5 821.7	3 175.5	468	255	5.8
1967	7 391.2	6 159.3	2 876.0	482	225	5.8
1968	7 819.9	6 516.6	3 042.8	496	232	5.8
1969	8 273.4	6 894.6	3 219.2	510	239	5.8
1970	10 838.2	9 031.8	4 217.2	650	304	31.0
1971	12 572.3	11 326.4	5 327.2	794	374	15.9
1972	14 583.9	13 201.6	6 179.6	901	422	16.0
1973	17 354.8	15 634.9	7 353.7	1040	489	18.9
1974	20 339.8	21 187.3	8 618.6	1374	559	17.1
1975	24 407.9	25 424.9	11 906.3	1603	751	20.0
1976	27 092.7	28 221.5	12 601.3	1735	775	10.9
1977	26 009.0	27 092.7	12 097.2	1624	725	−4.0
1978	30 430.5	32 720.9	16 360.5	1912	956	16.9
1979	34 995.0	41 660.8	19 550.3	2374	1114	14.9
1980	35 590.0	41 383.7	20 935.3	2295	1161	1.7
1981	36 479.7	39 651.8	20 610.0	2147	1116	2.5
1982	40 930.2	42 196.1	19 306.7	2229	1020	12.2
1983	45 923.7	45 023.3	21 065.9	2346	1042	12.2
1984	47 163.7	39 303.0	19 984.6	2002	1018	2.7
1985	48 437.1	45 268.3	19 933.0	2220	978	2.7
1986	49 454.3	48 484.6	22 176.8	2324	1063	2.1
1987	51 086.3	54 347.1	23 872.1	2544	1117	3.3
1988	52 618.8	55 977.5	25 056.6	2558	1145	3.0
1989	54 197.4	55 873.6	25 808.3	2481	1146	3.0
1990[e]	55 443.9	57 158.6	26 401.8	2233	1031	2.3

[a] In millions of North Korean *won*.

[b] Converted into US dollars by using North Korea's official exchange rate (see Table 2.11).

[c] Converted into US dollars by using North Korea's trade exchange rate (see Table 2.10). This estimate is still a little higher than the estimate made by the South Korean authority (National Unification Board). But it is not fair to intentionally underestimate North Korea's income by using the trade exchange rate instead of the official exchange rate. Instead, it should be argued that GNP or per capita GNP cannot be a measure of real living standards if much of it is going to military or heavy industry sectors which have nothing to do with consumer spending. Average income level is meaningless unless money income can buy consumer goods.

[d] This is GNP growth rate. Per capita GNP growth rate is calculated as the difference between gross GNP growth rate and population growth rate.

[e] Preliminary.

Sources: Tables 3.8 GVSP (I), and 3.10 (population).

TABLE 3.11(*b*) South Korea's GNP, per capita GNP, and foreign exchange rate (*won*/US dollars), 1953–1990

	GNP[a] (W billion)	GNP (US$ million)	Per capita GNP ($)	Growth rate[b] (%)	W/$ exchange rate[c]
1953	38.94	1 353	67		28.8
1956	121.98	1 450	66	−1.4	84.1
1960	210.71	1 948	79	1.1	108.2
1961	297.10	2 103	82	5.6	141.3
1962	348.90	2 315	87	2.2	150.7
1963	488.50	2 718	100	9.1	179.7
1964	700.20	2 876	103	9.6	243.5
1965	805.30	3 006	105	5.8	267.9
1966	1 032.50	3 671	125	12.7	281.3
1967	1 270.00	4 274	142	6.6	297.1
1968	1 598.00	5 226	169	11.3	305.8
1969	2 081.50	6 625	210	13.8	314.2
1970	2 776.90	8 105	252	7.6	342.6
1971	3 406.90	9 456	288	9.1	360.3
1972	4 177.50	10 632	318	5.3	392.9
1973	5 355.50	13 446	395	14.0	398.3
1974	7 564.50	18 701	540	8.5	404.5
1975	10 064.60	20 795	590	6.8	484.0
1976	13 818.20	28 550	797	13.4	484.0
1977	17 728.60	36 629	1008	10.7	484.0
1978	23 936.80	51 341	1392	11.0	466.2
1979	30 741.10	61 361	1640	7.0	501.0
1980	36 749.70	60 327	1589	−4.8	609.2
1981	45 528.10	66 238	1734	5.9	687.3
1982	52 182.30	71 300	1824	7.2	731.9
1983	61 722.30	79 500	2002	12.6	776.4
1984	70 083.90	87 000	2158	9.3	805.6
1985	78 088.40	89 695	2194	7.0	870.6
1986	90 598.70	102 789	2505	12.9	881.4
1987	106 024.40	128 921	3110	13.0	822.4
1988	126 230.50	172 776	4127	12.4	730.6
1989	141 794.40	211 200	4994	6.8	671.7
1990[d]	168 437.80	237 900	5569	9.0	708.1

[a] All data from 1981 are estimated according to the new System of National Accounting (SNA). Data before 1981 are based on old SNA.

[b] Growth rate of GNP up to 1970 are based on 1975 constant prices; thereafter it is based on 1985 constant prices.

[c] Calculated by dividing current GNP in billions of South Korean *won* by GNP in millions of US dollars.

[d] Preliminary.

Sources: Economic Planning Board, *Major Statistics of Korean Economy* (each year); Bank of Korea, *Economic Statistics Yearbook* (1990).

South Korea began to overtake the North from 1976 on, but in terms of per capita income, this did not occur until 1986, according to the official exchange-rate-based estimates.

It must be remembered, of course, that the population of North Korea is just about half of that of South Korea. The smaller population boosted per capita GNP in the North relatively higher up to 1986. However, Seoul's downgradings of North Korea's GNP usually provide inaccurate per capita GNP figures, while official Pyongyang claims are somewhat exaggerated, and often very vague. Parris H. Chang visited North Korea and the USSR in mid-1985 and wrote in *Newsweek* (8 July 1985) of a 'standard of living comparable to Eastern Europe', where per capita income was well over $2000 in 1980 even in the more backward states like Romania.[29] Thus, the evidence both of other studies and of such first-hand observers prove that our estimation of North Korea's GNP based on the official exchange rate is quite a fair one for comparison with South Korea. North Korea was ahead of South Korea in terms of per capita income until 1986, according to the official-rate value in Table 3.11(a). However, higher per capita income does not necessarily mean that North Korea had a superior real standard of living. If the North Korean economy was so heavily biased towards heavy industry and defence, much of which would produce goods used by the state and other heavy industry, this might merely contribute to raising absolute per capita income, which is simply an 'average' GNP divided by the number of total inhabitants of the country. This ballooned per capita income has, in fact, nothing much to do with consumer benefits. In other words, heavy and defence products can only balloon income accounts, and thus per capita income, numerically, but it implies little true economic value for consumers.

Actual per capita income, or disposable income, must be much lower than the average data-set can show. It is undeniable that the South has advanced more significantly than the North in aggregate dollar GNP since 1976. In 1989 South Korea is estimated to have had a GNP about three times greater and a per capita GNP about twice that of North Korea, even if we refer to the official exchange-rate-based GNP of the North. The greater advance in aggregate GNP of South Korea in recent years has been possible through the enhancement of its industrial sectors, its domestic and international transport systems, and world trade competitiveness.

Many observers, including those who have been in both North and South fairly recently, agree that the South appears to be ahead of the North, possibly because of its relatively high growth during the last two decades.

But the dark side of high economic growth in the South cannot be underestimated today. The social costs of such rapid growth are not disputed here: they include very low wage levels for many workers, particularly women and farmworkers, very long working hours, poor or non-existent social services (such as sick pay, unemployment benefits, pensions), high inflation (at times), inadequate housing, poor safety precautions in industry, weak labour laws, a high level of pollution, the devastation of rural communities, etc. Most of these negative factors operated up to the mid-1980s and some of them are still extant. They can be recognized as 'social costs' if and only if the people begin to feel that they have more than enough of all basic needs. Social costs cannot be costs if people are still in the vicious circle of poverty, but apparently this is no longer the case in South Korea.

3.3   REAL LIVING STANDARDS

By macroeconomic evidence, the North was still well ahead of the South until 1976—achieving, on average, roughly double the per capita output of some selected items, as shown in Table 3.12, not to mention per capita GNP as recorded in Table 3.11(*a*). After 1976, however, most sources in the West and quantitative data evidence tend to put South Korea ahead in real living standards, although it has to be made clear on what basis this claim rests. It is evident, though, that the per capita GNP in the South reached $2505 in 1986, while the most favourable estimate of per capita GNP in the North was about $2300.[30] By 1989, the South was leading the North in per capita GNP by more than twofold if the official exchange rate is the conversion factor, or by approximately fourfold if the trade rate is used instead.

Apart from per capita GNP and those selected items in Table 3.12, some other available information can be taken into consideration. First, let us compare food intake. In 1985 the World Bank estimated that of the North to be slightly higher, i.e. 130% of the daily calorie requirement per capita, compared with 125% for the

TABLE 3.12 Production of selected products, North and South Korea, 1970, 1976, and 1988[a]

	1970		1976		1988	
	North Korea	South Korea	North Korea	South Korea	North Korea	South Korea
Electric power (billions of kWh)	16.5	9.2	21.8	23.1	278.9	854.6
Coal (millions of tons)	27.5	12.4	39.5	16.4	40.7	24.3
Crude steel (millions of tons)	2.2	0.5	2.8	2.7	5.2	1.5
Fertilizer (nutrient content) (millions of tons)	0.3	0.6	0.6	0.8	3.5	3.7
Cement (millions of tons)	4.0	5.8	5.0	11.9	9.8	30.5
Textiles (excl. yarn) (millions of square metres)	418	329	450	936	660	700
Food (thousands of tons)	4306	6943	4481	8206	5210	7299
(Grains)	(4000)	(6160)	(4273)	(7535)		
Fish catch (millions of tons)	0.7	0.9	1.2	2.4	2.2	3.2
Machine tools (thousands)	10.0	7.5	24.0	8.4	30.0	55.0
Trucks (thousands)	4.0	5.5	10.0	19.5	30.0	200.0
Zinc (thousands of tons)	83	3	125	27	295	247
Lead (thousands of tons)	61	3	80	8	88	130
TV sets (millions)	negl.	0.1	negl.	2.3	0.24	12.71

[a] It must be remembered that the population of South Korea is about twice that of North Korea. Therefore, in terms of per capita output of most commodities in this table, the North had roughly double the per capita output of these items in 1976.

Sources: US CIA, Handbook of Economic Statistics (each issue); National Unification Board, Comparison of Economies between North and South Korea (December 1989).

South.[31] A second criterion is life expectancy. The *World Bank Atlas* (1989) gives the South figures of 60 and 70 in 1970 and 1988, respectively, compared with 60 and 69 for the North in those years.[32] The total fertility rate for the South is recorded as 4.2 in 1970 and 2.1 in 1988, while for the North it is 5.4 and 3.5 in the corresponding years.

With regard to education, the North enforces universal free 11-year compulsory education for all the school-age children and gives scholarships to students at universities, colleges, and training centres at all levels. The state covers expenses for both upbringing and education amounting to more than W15 800 ($16 120 in 1990 official prices) for a child in the period from his or her nursery and kindergarten days until graduation from college or university.[33] South Korea provides pupils with only six years of free primary school education, although it is planning to extend its compulsory education programme to nine years starting in 1993 if general consensus requires the extension. In 1989, the literacy rate of North Korea is reported to have been about 95%, compared with over 90% in the South.

North Korea claims to spend far more effort and money on health and to have a much higher ratio of doctors to population than the South (see Table 3.13).[34] But the North's definition of 'doctor' does not always make it clear whether he or she is a licensed doctor who has received a regular education at medical school.

Among other criteria, male and female income and employment, urban and rural standards of living, and equity may be compared. North Korea claims to be implementing the principle of equal pay for equal work and maintains that women have access to all jobs. Certainly, the 1946 Law on Equality of the Sexes seems to lay down the legal basis for such an approach.[35]

Equal rights are constitutionally guaranteed in the South, too. But in both North and South Korea, most top jobs in the civilian economy are occupied by men. This is particularly true in top jobs in government and industry.

In 1988, about 50.5% of North Korea's population were women. They make up about 48% of the total work-force, about 55.5% of the agricultural work-force, and about 45.5% of the industrial work-force. A comprehensive breakdown of women employees by sector gives figures of 20% of the mining and underground work-force,

TABLE 3.13 Major social indicators, North and South Korea, 1960 and 1988

	North Korea		South Korea	
	1960	1988	1960	1988
1. Population ('000)	10 789	21 877	24 994	42 380
2. Daily calorie supply per capita (kcal)	2 000[e]	3 232[a]	2 000[e]	2 907[a]
3. Life expectancy (%)	50[e]	69	50[e]	70
4. Total fertility (%)	7.8[e]	3.5	5.3	2.1
5. No. of students ('000)	2 528	8 043	4 912	11 060
Kindergartens: no. of students ('000)		3 500	15	405
no. of schools	N/A	60 000	325	8 030
Elementary schools: no. of students ('000)	957	1 753	3 855	4 820
no. of schools	4 145	4 760	5 264	6 403
Middle schools: no. of students ('000)	1 031		620	2 524
no. of schools	2 839	2 596	1 073	2 429
High schools[b]: no. of students ('000)	394	4 666	279	2 319
no. of schools	1 069		647	1 750
Colleges[c]: no. of students ('000)	146	494	143	1 292
no. of schools	76	244	66	234
6. Hospitals	447	2 510[d]	150	592
No. of beds	32 698		11 197	90 277
7. Clinics	4 364	5 414[d]	3 482	19 033[e]
8. Medical doctors per 10 000 persons	11.0	24.0[d]	3.1	10.0
9. No. of persons per passenger car	negl.	negl.	690.0[e]	37.5
10. Housing supply ratio	80.0[e]	74.0	82.5	69.4

[a] 1986 data from the *World Bank Atlas* (1989), 7.
[b] Includes vocational schools.
[c] Includes junior colleges, universities, and graduate schools.
[d] In 1982, there were 1531 general hospitals, 979 specialized hospitals, and 5414 clinics. There were 24 doctors and 130 hospital beds per 10 000 population in 1983.
[e] Estimates on basis of various sources.

Sources: North Korea's *Chungang Nyungam* (Central Statistical Yearbook), each issue; South Korea's Economic Planning Board, *Handbook of Korean Economy*, each issue. John Paxton (ed.), *The Statesman's Year-Book, 1989–1990* (New York: St Martin's Press, 1990).

15% of the heavy industry work-force, 70% of the light industry work-force, and 30% in the forestry and timber industry work-force. In the teaching profession, women make up 80% of the people's (elementary) schoolteachers, 35% of teachers in middle schools, 30% of those in vocational schools, and 15% in colleges and universities. In North Korea, all the working people receive benefits equally from the state regardless of age, sex, occupation, or religious belief according to the DPRK constitution. (See its article no. 62 regarding women's rights.)

The situation in the South is rather different. For a variety of complex reasons, a much smaller percentage of women than in the North are employed for remuneration, although women are increasingly in demand in the light industry sectors.[36] In a capitalist market equilibrium, the real wage is in theory equal to the marginal productivity of the worker. The ratio of wages for women is estimated to be around 45% of average male earnings except for a few jobs such as those in schools, government offices, and public institutions.

Urban and rural standards of living are hard to assess in the North. The North Korean regime claims that there exists no difference between the standard of living in urban and rural areas. 'In our country, the working popular masses [both in urban and rural areas] are fully provided with equal liberties and rights as masters of the state and society, and all social wealth is geared to the promotion of their welfare.'[37] As such, the said egalitarian society must not permit the existence of any gaps in living conditions between urban and rural areas. But some Western visitors who have been to the North recently relate that rural living standards are still somewhat lower, as seems probable. Suffice it to say that, either from direct visual observation or from indirect hearsay, which is the most readily available source, it seems that there must be some degree of disparity of income and living standards among different types of workers and different regions in the North. However, the disparity is likely to be much less in the North than in the South. This can be easily checked if we compare the living conditions and monthly income and expenditure per household in both countries.

3.3.1 Socialism in the North

First, we shall consider the living conditions and welfare in North Korea. Just as the means of production are owned completely by the people, so is the living environment. The state gives out cereal rations bi-weekly, work clothes and ordinary clothing twice a year, and other knitted garments every quarter. Of course, people can buy their basic necessities both in state or co-operative stores and in the secondary private market in limited quantities, if necessary.

All housing units are either state-owned or in co-operative owner-ship. Dwellings are allocated to individual families according to the order in which they are ranked by the administrative author-ity. Most dwellings are either flats or (rented) terraced houses. The living space per general worker family is 26.4 m^2 (8 pyong), and this is equivalent to 5.3 m^2 (1.6 pyong) per individual.

Food too is either rationed by the government or distributed in rural areas according to each member's work contribution by the farm co-operative of which each farmer is a member. The idea behind food rationing is that, with an adequate provision of food for every member of society, people will cease to wish to improve the quality of their lives. They will come to regard food as nothing but a necessity for survival, and to see Comrade Kim Il-Sung as their divine leader and provider.

Aspects of consumerism as well as social life have almost similar features in all communist countries. North Korea's leaders claim that all social wealth is always geared to the promotion of the material welfare of the working people. Throughout the imple-mentation of its several economic plans, Kim has often emphasized that his *chuche*-oriented policy aims to enhance the living condi-tions of the masses.[38] Nevertheless, the '*chuche* idea' has now become more difficult to promulgate. Perhaps it is the sorry fate of a communist country for its policy-makers to adhere to the view that the route to an advanced, affluent economy is through heavy industry, while paying lip-service to the material well-being of the consumers. Because the greater proportion of production is designated for use by the military (to be further discussed in the next chapter), the people remain deprived of consumer goods.

Yet, there is much less disparity in incomes in the North, with at most a differential of fivefold between the rich and the poor. Similarly, there is not likely to exist any marked inequality in wealth distribution. However, it *is* likely that there is a visible élite in the DPRK made up of Workers' Party and government officials who enjoy some forms of conspicuous consumption in non-earned services provided by the state. Table 3.14 shows monthly wage earnings by groups, Table 3.15 gives daily cereals rationing standards, and Table 3.16 illustrates the composition of monthly consumption expenditure per family. As is shown in Table 3.16, every family in North Korea appears to save a lot monthly, given the seemingly low level of incomes. This may be possible

TABLE 3.14 Monthly incomes per worker, North Korea, 1985 and 1989

	Monthly wage (W)[a] 1985	1989	Remarks
Office workers			
High-ranking officials	300–350	300–350	High-ranking WPK (Workers Party of Korea) officials and cabinet members
Local officials	100–150	100–150	Province and county chief officials
General officers	250–490	250–490	
Trainee officers	70	70	
Technicians			
Chief technicians	150–200	150–200	First-class enterprise and factory managers
General technicians	75–78	75–100	Grade 5 technicians (Grade 8 is highest)
Physical workers			
Heavy workers	130	130	Mineworkers, steel factory workers
Light workers	90	90	General machine tool operators
Average workers	70	70	Very light work
Sailors (ocean)	300–350	300–350	
Teachers			
College/university	200–250	220–300	
General teachers	100–220	220	
Military officers			
Generals (commander)	250–490	250–490	
Colonels	120–220	120–220	
Lieutenants	80–110	80–110	
Others			
Medical doctors	95–250	95–300	
Actors	200–300	200–300	People's Actors and Awarded Actors

[a] The North Korean *won* was officially exchanged at approximately a 1 : 1 ratio with the US dollar during the 1980s; but its trade exchange rate is around W2.36 = $1, while the non-trade (tourist) rate is estimated at W1.66 = $1. Also, if average monthly income per person is about W200, then the annual income per capita would be W2400, which was about $2400 at the official exchange rate in 1989.

Sources: visitors to North Korea (see Yu Eui-Young, 'North Korean Students, People, and Religious Leaders Whom I Met', *National Unification* (March–April 1990)), and data from the National Unification Board, Seoul.

TABLE 3.15 Rationing standards of daily cereals, North Korea, 1989[a]

	Quantity (g)	Food mix ratio (Rice : misc. grain)
Party and government officials	700	10 : 0
Special mission soldiers	800	7 : 3
Soldiers	700	3 : 7
Heavy workers	800	3 : 7
General workers	600	3 : 7
College students	600	3 : 7
Highschool students	500	3 : 7
Primary school pupils	400	3 : 7
Pre-school children	300–100	3 : 7
Other dependants	300	3 : 7

[a] Except for grain, most side-dishes must be purchased at the food store. But soy sauce, beanmash, red-peppermash, and salt are also rationed.

Source: National Unification Board (Seoul), *Survey of North Korean Economy* (1990).

TABLE 3.16 Monthly expenditure per household in North Korea, by income group, 1984 (%)

	Assistant managers	Grade 8 workers	Clerks
Cereals and side dishes	17.6	18.0	21.1
Clothing and footwear	27.8	19.3	22.6
Housing and fuel, light	6.0	3.0	3.0
Education expense	2.1	2.3	2.7
Culture and sanitary	5.6	8.6	10.1
Public fees	1.0	1.0	1.0
'Carry-over' and others[a]	39.9	47.2	39.5

[a] 'Carry-over' is considered to be mostly family savings. In North Korea, a high rate of saving seems possible, partly because consumers are constrained from spending with the limited availability of commodities and partly because prices are kept low by the Price Determination Bureau of the State Planning Committee.

Source: Institute of North Korea Study *North Korea Survey* (Seoul, 1983–1985, 1986 Supplement), 76.

partly because most of the food, clothing, and shelter are supplied free by either the state or a co-operative, and partly because the Price Determination Bureau of the State Planning Committee sets all prices low.[39] It may also reflect the lack of things to buy in the limited market, as it does in other communist bloc countries.

According to most observers who have been to North Korea in 1989 and 1990, the price tags of some commodities at stores in Chollima ('Flying Horse') Street of Pyongyang show that 1 kg of sesame seeds costs W1, a pair of socks W1, a pair of tennis shoes W5, a woman's sweater W7, a pair of men's shoes W30, and an aluminium cooking pot W20. The price of a book ranges from 50 *chon* (that is, W0.5) to W7.50. The underground fare (one way) in Pyongyang City is just 2 *chon* (W0.02). At Pyongyang railway station, single ticket prices are W3.30 to Shineuiju, W4.40 to Hamheung, and W9.15 to Chungjin. A discount ticket window is available for military veterans, journalists, and schoolteachers.

Urban workers usually work six days a week with 15 days' paid vacation a year. Rural farmers have every first, eleventh, and twenty-first day of the month off. Every office worker without exception is assigned one day of manual (physical) work a week. Most white-collar workers (with the exception of those in the teaching profession) must engage in some sort of manual labour such as digging ditches every Friday. Saturday is 'education' (indoctrination) day for all people.

On average, men marry around the age of 27 while women are married at about 24. Love matches are quite common today, with parents giving their approval for such marriages. The average cost for marriage is estimated to amount to W1500–2000 on each side for gifts and living preparations, etc. Because of the limited availability of housing for new couples, they usually live either with one of their parents or in a relative's house until they are allocated a dwelling. Retirement ages are 55 for women and 60 for men. There is no age limit for professors in colleges and universities. After retirement, about 60% of the last monthly salary is paid as a pension.

3.3.2 Capitalism in the South

Now let us look at living conditions and income distribution in the South. Per capita GNP in 1983 was estimated at almost exactly

$2000, up from $80 in 1961. In 1990 it reached more than $5000. South Korea and its admirers describe this achievement, coupled with the hosting of the 1988 Olympic Games in Seoul, as the 'miracle of the Han River' and 'a second Japan'. North Korea, on the other hand, and critics of Seoul claim that the South's economy is a 'house built on sand', and even that the South's industrialization is 'fictitious'.[40] Many criticisms of the South can be raised in terms of macroeconomic or political perspectives such as its foreign dependency on food and defence and the deteriorating natural environment which is accompanying the process of rapid industrialization. But here discussion will be strictly confined to some micro aspects, namely living conditions and income distribution, so as to pave the way for comparison with the North.

On the whole, the South is a capitalist free-market society, although there are still some shortcomings in the field of human rights and the level of democratization. There is no longer free food, clothing, and shelter from 'helicopters' (government or higher authorities), except for marginal cases of disabled people with no dependants who badly need government care and aid. Demand for and supply of goods principally depend on the invisible hand of the market. Very often, however, the inadequacies of short-sighted government policy can result in serious problems and social externalities. For instance, income and housing policies pursued by the South Korean government are intended to have an equalizing impact on the distribution of income and wealth. Nevertheless, such government programmes have often been misguided; and, together with some other economic and social factors, market intervention frequently turns out to act as a disequalizing factor in the distribution of wealth and income. Today, there exists a growing inequality in the distribution of wealth, and burning class conflicts between the haves and the have-nots. Before proceeding further into this matter, we shall briefly discuss the way a household obtains shelter in the market economy so as to enable comparison with the allocation of free shelter in the North.

Housing in South Korea usually takes the form of private ownership. An individual family buys its own dwelling out of its own savings, loan, or inheritance. If it cannot afford to buy a place in which to live, it has to rent one, on either a monthly payment basis or a longer contract arrangement, in the private housing market. Mortgage loan markets have not fully developed in South Korea

TABLE 3.17 South Korea's investment in the housing sector and the price index of major building materials, 1970–1987

	1970	1975	1980	1985	1987
Housing investment rate	3.1	4.5	5.9	4.8	5.0
Price index of major building materials (1985 = 100)	15.9	33.6	80.6	100.0	101.5

Source: Economic Planning Board, 'Population and Housing Census', *National Income Accounts* (Seoul: Bank of Korea, 1989).

yet except for one- to three-year mortgage loan arrangements. During the two decades of rapid economic expansion, the loan market has always been on the 'short side', in that excess demand for loans constantly prevails in this rapidly growing economy.

The demand for housing has exploded, exceeding the supply, partly because the number of households has grown more rapidly than expected, owing to the increase of nuclear families. This excess demand is exacerbated by the growing trend in property speculation, because many investors are looking for windfall capital gains in this overpopulated country. The cost of building materials has risen, and the continuing government regulation of building permits and offer prices of new blocks of flats has been strict. These economic and social factors have contributed to a fall in the house ownership rate, as shown in Table 3.13: the rate was 68.2% in 1970, down to 58.4% in 1980, and still further to 53.5% in 1985, but it rose to 69.4% in 1988 because of somewhat accelerated efforts to increase the supply (see Table 3.17).

The housing (floor) space per household (national average) was about 35.9 m² in 1970, 41.4 m² in 1975, 45.8 m² in 1980, and 45.3 m² in 1985. The living space per capita is 6.8 m², 8.2 m², 10.1 m², and 11.1 m² in those respective years, which indicates that floor space has become greater over time. By region, the floor space is smaller than the national average in urban areas, indicating the intense population density, and greater in rural areas. It is the same in the North.

Methods of financing a house purchase and the average period necessary for buying a house vary. It is noted that family savings account for the largest share of housing finance, and that the length of time is highly centred around the six- to nine-year period. The proportion of unmarried people who buy a house is only 19.9% of total unmarried adults, much lower than that by married people (80.1%), as the national average indicates.

With regard to urban and rural standards of living in the South, no one is claiming that rural living standards are the same as those in urban areas. But the difference is difficult to evaluate in terms of either per-household income or per-capita consumption expenditure data, because the average figures have moved upward numerically in favour of rural areas as more people migrate into urban areas. The percentage of farm population (and households) in South Korea has dramatically decreased, from 49.4% (49.0%) in 1969 to 44.7% (44.5%) in 1970, to 20.9% (20.1%) in 1985, and again to 17.3% in 1988. Thus, cultivated area per farm household has increased, from 0.91 ha (0.50 ha for paddy and 0.4 ha for upland) in 1967 to 1.17 ha (0.74 ha paddy, 0.43 ha upland) in 1988.

The ratio of farm household income and consumption expenditure is shown in Table 3.18. Generally speaking, the rural family seems to enjoy relatively higher income and consumption expenditure than the urban family, perhaps because of a higher ratio of resources per family in sparse rural areas, particularly since the early 1980s.

A brief look at the composition of average monthly income and consumption expenditure per salary and per wage-earner's family by income quintile shows that the proportion of total expenditure going on food and beverages drops in both urban and rural areas as incomes rise, reflecting Engels' Law. Demand for furniture and utensils, clothing, health care, education, and other such expenses is seen to be more income-elastic, as the theory predicts.

Now we need to consider the question of equity in regard to income (flow variable) and wealth (stock variable) distribution in the South.

Income inequality in South Korea has improved slightly over the last decade, but wealth inequality has worsened significantly. The general feeling is that the drive for rapid economic growth in the South has widened the gap in wealth accumulation among different classes, although the conventional Gini coefficient (the coefficient of concentration)[41] in both wage income and non-property income has steadily decreased over time. When data are used to support this contention,[42] it is observed that the degrees

TABLE 3.18 Ratios of rural household income and consumption expenditure to urban household income and consumption expenditure in South Korea, 1965–1989 (in *won*)

	Urban households		Rural households			
	Monthly income (A)	Consumption expenditure (B)	Monthly income (C)	Consumption expenditure (D)	C/A (%)	D/B (%)
1965	9 380		9 350		99.7	
1970	28 180		21 317		75.6	
1975	65 540	58 350	72 744		111.0	
1976	88 270	72 650	96 355		109.2	
1977	105 910	84 190	119 401	81 367	112.7	96.6
1978	144 510	112 680	157 016	110 042	108.7	97.7
1979	194 749	149 142	185 624	138 514	95.3	92.9
1980	234 086	180 531	224 426	178 193	95.9	98.7
1981	280 953	216 648	307 321	223 007	109.4	102.9
1982	317 052	251 489	372 098	271 486	117.4	108.0
1983	364 019	274 607	427 354	337 806	117.4	123.0
1984	402 297	299 392	462 428	356 018	114.9	118.9
1985	431 183	322 558	478 021	390 904	110.9	121.2
1986	481 018	353 956	499 584	416 225	103.9	117.6
1987	561 675	406 955	544 610	433 387	97.0	106.5
1988	646 672	467 639	677 468	502 554	104.3	107.5
1989	804 938	594 287	n.a.	n.a.		

Sources: Economic Planning Board, *Family Income and Expenditure Survey*; Ministry of Agriculture, Forestry and Fisheries, *Farm Household Economic Survey* (each year).

of inequality are greatest for property income, then total income (including the earnings flowing out of property such as physical assets as well as financial assets), non-property income, and wage income, as shown in Table 3.19. From the Gini coefficients, estimated from the urban family income survey data (1979–87) by income decile, we can derive the following observations.

1. Inequality in income from wealth is much higher than inequality in total income including the income earned from properties. The Gini coefficient of the property (wealth) income in 1987 was 0.487 compared with 0.289 for the Gini coefficient of total income. The total income index has improved relatively,

TABLE 3.19 The Gini index of non-farm household income, South Korea, 1979–1987

	Wage income	Income excl. property earnings	Total income	Property earnings	Rent	Interest and profits
	(A)	(B)	(C = A + B + D)	(D = E + F)	(E)	(F)
1979	0.289	0.295	0.298	0.387	0.345	0.425
1980	0.278	0.286	0.289	0.378	0.263	0.453
1981	0.271	0.286	0.290	0.379	0.207	0.472
1982	0.280	0.289	0.294	0.456	0.449	0.464
1983	0.277	0.288	0.293	0.450	0.449	0.451
1984	0.274	0.285	0.291	0.480	0.422	0.540
1985	0.280	0.293	0.299	0.491	0.487	0.497
1986	0.269	0.283	0.289	0.457	0.464	0.445
1987	0.267	0.279	0.289	0.487	0.474	0.503

Source: EPB, *Annual Report on the Family Income and Expenditure Survey* (each year). For further reference, see Lee Hak-Yong, Hwang Eui-Gak, and Kim Dong-Won, 'A Study on Improvement of Financial System for Change in Wealth Distribution' (in Korean), unpublished report submitted to Korea Development Institute, November 1988.

perhaps as a result in the rise in the share of wage income going to the bottom 20% of the population (the bottom quintile), from 8.3% in 1979 to 8.8% in 1987, while the amount received by the top quintile, about 38.0%, has decreased somewhat over time (see Table 3.20). Social trends such as lower fertility rates and increasing demand for light industry workers have increased female labour force participation to raise family incomes slightly in the lower and middle parts of the income spectrum.

2. Income inequality has steadily improved, from 0.298 in 1979 to 0.289 in 1987; but the Gini index in property income has deteriorated significantly, from 0.387 to 0.487. Of the property income index, the analysis shows that inequality is greater in earnings from financial assets (such as interest and profits) than in income from physical assets (only rent, in this analysis). This is probably because other income from physical assets, such as turnover gains (capital gains) from housing and land transactions, are not included—because such information is not given in the official publications. Of course, income from financial assets must have been

TABLE 3.20 Share of wage income by income decile, South Korea, 1979–1987

	I	II	III	IV	V	VI	VII	VIII	IX	X	Total
1979	0.034	0.049	0.061	0.071	0.080	0.091	0.105	0.123	0.151	0.235	1.00
1980	0.034	0.052	0.061	0.071	0.082	0.092	0.105	0.123	0.150	0.229	1.00
1981	0.035	0.051	0.062	0.071	0.081	0.093	0.106	0.123	0.150	0.229	1.00
1982	0.035	0.050	0.061	0.071	0.081	0.092	0.105	0.123	0.150	0.233	1.00
1983	0.036	0.051	0.061	0.070	0.080	0.091	0.105	0.123	0.149	0.234	1.00
1984	0.035	0.051	0.062	0.071	0.082	0.092	0.105	0.120	0.148	0.233	1.00
1985	0.034	0.050	0.060	0.070	0.081	0.092	0.105	0.122	0.150	0.237	1.00
1986	0.035	0.051	0.061	0.071	0.082	0.092	0.106	0.123	0.150	0.228	1.00
1987	0.036	0.052	0.062	0.071	0.081	0.092	0.106	0.122	0.150	0.228	1.00

Source: as for Table 3.19.

TABLE 3.21 Ratios of property income[a] to total income by income decile, South Korea, 1979–1987

	I	II	III	IV	V	VI	VII	VIII	IX	X
1979	2.37	2.66	3.09	2.68	2.60	2.81	2.63	2.87	3.27	4.48
1980	1.73	2.73	2.63	3.82	3.51	3.28	4.14	3.96	4.52	4.40
1981	1.31	2.95	3.39	3.21	4.43	4.23	5.46	5.41	4.22	5.02
1982	1.45	1.94	1.97	1.80	1.35	2.56	2.86	2.74	3.57	4.19
1983	1.45	1.03	1.35	2.11	2.70	2.21	2.63	3.29	4.26	3.48
1984	1.76	1.19	2.42	2.04	2.66	1.81	1.84	2.80	3.32	5.29
1985	1.84	1.49	1.65	2.00	1.85	2.14	2.22	3.24	4.52	4.56
1986	1.86	1.97	1.57	1.99	2.57	2.22	3.48	4.04	4.35	4.61
1987	0.98	1.35	2.19	2.22	2.70	2.93	3.80	3.95	4.57	5.74

[a] Sum of interest earnings, profit share, and rent earnings.

Source: as for Table 3.19.

very high in view of the sustained high rates of interest, returns on capital, and the supposedly large scale of secondary (underground) markets for money that has existed during the last couple of decades in South Korea.

3. The percentage of property income to total income by income group shows that the share going to the bottom 10% of the sample population was 2.37% while that going to the top 10% was 4.48% in 1979; but in 1987 the shares changed to 0.98% and 5.74%, respectively. Such a significant change indicates that the wage income share increased relatively more for the lower-income group, while for the higher-income group it was due either to an increased share of wealth (property) ownership or to a relative rise in the rate of return from assets, or both (see Table 3.21).

Thus, the dis-equalizing tendency of overall income distribution seems to lie in the significant increase in the inequality of property income, even though the distribution of wage income is improving considerably. The inequality of property income is related either to an unequal distribution of wealth ownership or to a relatively faster growth in the profit rate. The gap would be much wider if various capital gains (to be made in property transactions without much tax burden) other than rent earnings were included in the sample. Admitting that the Gini index of total income is steadily improving over time (and the Korean case is not so bad, compared with indexes in many other market economies[43]), the degree of psychological or relative deprivation being felt by those people who regard themselves as 'below the median' has certainly widened recently in the South. This feeling of alienation occurs partly because of the sudden emergence of 'rich neighbours', taking advantage of all opportunities (including information on future development sites and inflation expectation) and enjoying various favours from government in the stages of economic take-off.

A measure of inequality must be relative. The question is most closely related to equality, fairness, impartiality, and justice in all aspects of human activities including wealth (and income) distribution, the production system, and social institutions, which are truly subjects beyond this study.

Although it is hard to compare the relative spread of feelings of inequality (or relative deprivation) between an upper-income stratum and those below the median in both countries, there would seem to be much more disparity of wealth and income and in consumption in the South than in the North. This is likely to be grounded in the natures of the two different economic systems.

3.4 CONCLUSION

In this chapter, we have begun to consider the problems of estimating a comparable GNP for North Korea on the basis of a combination of almost non-existent statistics and inadequate information. However, the derivation of three alternative GNPs from the GVSP estimate and other sources was attempted and tested by means of the principal components analysis. The chosen measure was then converted into US dollar values by using both the North's official exchange rate and its trade exchange rate to make it comparable with South Korea's GNP in aggregate and per capita figures. When the official exchange rate is applied, the North is seen constantly to lead the South in per capita GNP until 1985. If the comparison is based on the trade exchange rate (which is more than twice the official rate), the South is seen to have overtaken the North in per capita GNP in 1974. Meanwhile, Seoul claims that it overtook the North in 1969, while the US CIA claims that this happened in 1976.

It is undeniable that the South advanced further than the North in the 1980s, but it seems unfair to use the trade rate instead of the official rate to ludicrously undercut the North's actual GNP. Perhaps it is more reasonable for the North's critics to argue that GNP or per capita GNP has nothing much by which to measure real living standards in a socialist command economy where heavy industry (and the hidden defence sector) takes such a great portion of GNP. Thus, per capita GNP, which is nothing but a statistical average, does not indicate the level of private consumption, just as a high saving ratio in any communist country has nothing to do with an increase in capital stocks or future spending. (Income is being nominally saved only because there is nothing to spend it on in an extremely limited market for consumer goods.) Average per capita income data are meaningless where the large part of GNP goes to heavy industry and government and where there is no market supply of consumer goods for the individual's money income to chase after in that closed economy.

Nevertheless, people in the North appear to have a higher standard of living, with well-balanced ratios from the state comparable to Eastern Europe, where per capita income was well over $2000 by 1980. This allows a rough inference that North Korea's claims to have achieved a per capita income of $1920 in 1979 does not deviate much from the actual standard of consumption corresponding to that level of income. The level of per capita GNP can also be inferred from the data on average monthly earnings per capita in 1985 and 1989, which were collected from almost all observers who had been to North Korea (as seen in Table 3.14). For instance, if average monthly income per head were W200 (North Korean currency) in 1985, an annual income of W2400 would result. That income corresponded to an almost equivalent amount in US dollars in 1985. Thus, in 1979, six years earlier than 1985, a per capita income of $1920 does not seem to have been 'manipulated' or to deviate from the true line at all. This fact, in turn, shows that the official exchange rate is a more appropriate conversion factor than the trade exchange rate in evaluating North Korea's GNP in US dollar terms.

The comparison of per capita GNP between the North and the South, then, shows that the South moved ahead of the North only in 1986 (see Tables 3.11(*a*) and 3.11(*b*)). In 1989 the South's GNP per capita was about twice that of the North, showing that the North's economy has stagnated very much since the early 1980s.

Such evidence as exists of consumers' living patterns and standards appears to reveal that those in the South seem to be higher than in the North, particularly in the 1980s, even if it also has to be said that the South's superiority is hardly demonstrated by the available statistical data. The average standard of living, not to mention freedom of the individual (which is *the* most valuable criterion), seems to be superior in the South, but wealth and income are less evenly distributed among different groups.

Other indicators, such as life expectancy, mortality, literacy, and other physical quality-of-life indexes, merely serve to indicate that there exists no significant difference between the North and the South.

NOTES TO CHAPTER 3

1. Nikolai Shmelev and Vladimir Popov, *The Turning Point: Revitalizing the Soviet Economy* (New York: Doubleday, 1989), 76.
2. The statement that NMP covers only the production of goods and excludes services is not fully accurate. First, freight transport and material communications and trade are services that are always in NMP. Second, in moving from the gross output of any sector of material production to net product, the material product system subtracts material inputs and depreciation but leaves in the value of services purchased from firms in the non-material sphere—such as research and development, banking, insurance, education, and health—which need to be removed because they will be included in the estimates of the value added in the non-material sectors. Third, social services, provided free of charge to employees in the material sectors, such as health care and reimbursement for business travel, are also left in as components of NMP.
3. Transaction revenue (profits in turnover) is equivalent to excise (sales) taxes, or the difference between retail price and wholesale price in a market economy.
4. Paul Marer, *Dollar GNPs of the USSR and Eastern Europe* (Baltimore: Johns Hopkins University Press for World Bank, 1985), 20.
5. This is the source of chronic excess demand in the consumer goods market.
6. Marer, *Dollar GNPs*, 12. For various estimates of North Korea's GNP, see Kyong-Mann Jeon, 'Differences in Estimates of North Korea's GNP', Working Paper, Rand Corporation, Santa Monica, Cal., March 1982.
7. Harrison Salisbury, *To Peking and Beyond: A Report on the New Asia* (New York: Quadrangle, 1973), 199, 200. See also Jon Halliday, 'The Economics of North and South Korea', in John Sullivan and Robert Foss (eds.), *Two Koreas— One Future?* (University Press of America, 1989), 27–8.
8. René Dumont, *The Hungry Future* (London: Deutsch, 1969), 137; and Joan Robinson, 'Korean Miracle', *Monthly Review* (London) (January 1965); both cited in Halliday, 'Economics', 27–8.
9. Halliday, 'Economics', 30. See also 'Moscow', *New York Times* (August 1984), 13.
10. John Merrill, 'North Korea's Halting Efforts at Economic Reform', paper

presented at the Fourth Conference on North Korea, Seoul, 7–11 August 1989, 3.

11. J. R. Hicks, 'Income', in R. H. Parker and G. C. Harcourt (eds.), *Readings in the Concept and Measurement of Income* (Cambridge University Press, 1969), 81–2.

12. Institute of Economics of Academy of Social Sciences of DPRK, *Economic Dictionary 2* (in Korean) (Pyongyang: Social Science Publishing Co., 1973), 134; see also the 1970 edn., 219–20.

13. Joo-Whan Choi, 'Estimates of North Korea's GNP', *The Unification Policy* (in Korean), 4:1 (1978); Young-Kyu Kim, 'Methods of North Korea's GNP Estimates', *The Unification Policy* (Seoul), 6: 3–4 (1980); Poong Lee, *Methods of GNP Estimates of North Korea* (Seoul: North Korea Research Institute, 1981); Ha-Chung Yeon, *North Korea's Economic Policy and Management* (Seoul: Korea Development Institute, 1986), 131–43; Myung-Kyun Cho, *Explanation of North Korea's GNP Estimation Approaches* (Seoul: National Unification Board, 1988), 1–51; NUB, 'The Survey of North Korea's Economy', each issue.

14. See the *World Bank Atlas*, each issue before 1985; US CIA, *Handbook of Economic Statistics*, each issue; US ACDA, *World Military Expenditures and Arms Transfers* (1985); IISS, *Yearbook, 1987–8*; SIPRI, *Yearbook 1987*; EIU, *Country Report: China and North Korea* (quarterly); FSO, *Statistik Des Auslandes: Korea, Demokratische Volksrepublik* (1986). See also Pong S. Lee, 'An Estimate of North Korea's National Income', *Asian Survey*, 12:6 (University of California Press, 1972), 518–21; and Joseph S. Chung, 'A Study of North Korean Economy' (in Korean), *Korea and International Politics*, 5: 2 (Kyungnam University, 1989).

15. For a brief summary of differing methods of some estimates, see Yeon, *North Korea's Economic Policy*, 120–43, and Myung-Kyun Cho, *Explanation*, 6–51. See also Kyong-Mann Jeon, *Differences in Estimates of North Korea's GNP* (Santa Monica, Cal.: Rand Corporation, 1982).

16. Most of the sources do not fully explain the methods used.

17. Marer, *Dollar GNPs*, 15–16 and 27–119.

18. See also *North Korea's Statistics, 1946–85* (Seoul: National Unification Board, 1986), 124–7. North Korea's GVSPs (1946–65) in North Korean currency are given there and reproduced in Table 3.3.

19. Joseph S. Chung, *The North Korean Economy: Structure and Development* (Stanford, Cal.: Hoover Institution Press, 1974), 248. See also Joseph S. Chung, 'The Six-Year Plan of North Korea: Targets, Problems and Prospects', *Journal of Korean Affairs* (Seoul), 1: 2 (1971), 21.

20. *FAO Production Yearbook* (1985), *Monthly Bulletin of Statistics* (December 1987), and *Vantage Point* (each issue); requoted from Won-June Lee, 'North Korea's Agricultural Policy', *North Korea* (Seoul: North Korea Study Institute, 1982). See also National Unification Board, *A Study of North Korea's Agricultural Production* (Seoul: NUB, 1989), 17.

21. This growth rate is grounded only on the production of principal crops, excluding livestock. Data used to calculate the growth rate are from Economist Intelligence Unit, *Country Reports: China and North Korea* (1988–9), 60.

22. See *Asia Yearbook* (Hong Kong: Far East Economic Review, 1984). But the ground for this estimation is not provided.

23. The PPP gives the number of units of a country's currency that have the same purchasing power for the category as a comparable country's currency. For a binary comparison between a partner country and the USA the purchasing power parity for each detailed category is calculated as the geometric mean (usually unweighted because, in the absence of expenditure data at the item

level, no estimate of the relative importance of different items is available) of the price ratios for the item. See also Thomas Wolf, 'Exchange Rates, Foreign Trade Accounting, and Purchasing Power Parity for Centrally Planned Economies', World Bank Staff Working Papers no. 779 (1985).

24. Before reducing the GVSP estimate to a GNP concept, the principal components analysis is attempted in order to choose the best estimate. For a statistical methodology of principal components analysis, see Phoebus Phrymea, *Econometrics: Stratified Foundations and Applications* (New York: Harper & Row, 1970), 53–65. Suk-Bum Yoon had also attempted to calculate the principal components from five series of North Korea's GNP estimates (1960–84) made by South Korean scholars. But the data for North Korea's GNPs, which Yoon used, appear to have been underestimated, not to mention that their estimation methods were too arbitrary. (See Suk-Bum Yoon, 'A Trial Estimation of An Econometric Model of North Korea', paper prepared for the International Symposium on Inter-Korean Economic Cooperation, Seoul, 28–9 August 1989.)

25. The results of multiple regression are as follows:

$$X_1 = 331.6151 + 0.8366X_2 + 0.0124X_3 \qquad R^2 = 0.9999$$
$$ (3.549) \quad\ (74.976) \quad (1.117)$$

$$X_2 = -381.2765 + 1.1888X_1 - 0.0093X_3 \qquad R^2 = 0.9987$$
$$ (-3.375) \quad\ (74.976) \quad (-0.698)$$

$$X_3 = -2370.6122 + 3.1221X_1 - 1.6570X_2 \qquad R^2 = 0.9766$$
$$ (-1.389) \quad\ (1.117) \quad (-0.698)$$

$$Y_1 = 1348.6958 + 0.1778Y_2 + 0.6637Y_3 \qquad R^2 = 0.9812$$
$$ (1.140) \quad\ (2.641) \quad (10.013)$$

$$Y_2 = 3496.5978 + 1.0331Y_1 + 0.0781Y_3 \qquad R^2 = 0.9206$$
$$ (1.230) \quad\ (2.641) \quad (0.238)$$

$$Y_3 = -2003.2992 + 1.1509Y_1 + 0.0233Y_2 \qquad R^2 = 0.9770$$
$$ (-1.294) \quad\ (10.013) \quad (0.238)$$

Figures in parentheses are t-values.

26. The estimation results of correlation between each GVSP data series and its first principal components (PC) are as follows:

$$X_1 = 34934 + 16778\ PCX_1 \qquad R^2 = 0.9974;\ RMSE = 1505.9955$$
$$ (135.259) \ (110.548)$$

$$X_2 = 40781 - 14169\ PCX_2 \qquad R^2 = 0.0028;\ RMSE = 34663.8363$$
$$ (6.860) \ (-0.298)$$

$$X_3 = 39122 + 295\ PCX_3 \qquad R^2 = 0.0000;\ RMSE = 34910.3656$$
$$ (6.543) \ (0.0000)$$

$$Y_1 = 35442 + 17912\ PCY_1 \qquad R^2 = 0.9881;\ RMSE = 3424.1088$$
$$ (60.355) \ (51.458)$$

$$Y_2 = 43223 + 29452\ PCY_2 \qquad R^2 = 0.0378;\ RMSE = 36046.9538$$
$$ (6.992) \ (1.120)$$

$$Y_3 = 39794 + 24082\ PCY_3 \qquad R^2 = 0.0047;\ RMSE = 37228.9960$$
$$ (6.233) \ (0.388)$$

27. According to the 'law of equal cheating', all relative data such as growth figures are not affected as long as the extent of falsification, omission, errors, etc., remains constant, and this is assumed to be so for both countries.

28. Halliday, 'Economics', 42.

29. Ibid. 42.

30. Most South Korean sources give per capita income in the North as about half that in the South in 1986. This would be so if the trade exchange rate is used to convert North Korean *won*-denominated GNP into US dollars, instead of the official exchange rate.

31. *World Development Report* (1985).

32. The North disputes these figures and claims an average life expectancy of 73 in 1976, which breaks down into 71 for males and 77 for females in 1988.

33. Pyongyang KCNA, 21 March 1990. 'Many Popular Welfare Measures in Force in North Korea': see FBIS-EAS-90-056, 22 March 1990, 20–1.

34. North Korea's Major Measures for People's Health Programmes:

(1) February 1946	Health Bureau of DPRK established	
(2) 20 March 1946	Guidelines laid down to provide free medical care for poor people	
(3) January 1947	Social insurance system introduced	
(4) 30 July 1950	Free medical treatment law for war casualties promulgated	
(5) 15 October 1950	Complete free medical care system introduced at the Central Workers' Party's 15th Meeting (cabinet order no. 203)	
(6) June 1958	Academy of Medical Science established	
(7) 1969	Doctors' coverage area assigned	
(8) 1970	All rural area clinic facilities expanded to hospital levels	
(9) 3 April 1980	People's Public Health Law of DPRK promulgated	

35. Halliday, 'Economics', 39–40.

36. In the South, the female labour force participation rate was 37.2% in 1965, 40.4% in 1975, and 45.0% in 1988. The percentage of female teachers to total teachers in South Korea was 47.3% in primary schools, 42.1% in middle schools, 20.6% and 22.6% in general and vocational high schools respectively, and 16.5% in colleges and universities in 1988. For details, refer to Economic Planning Board (EPB), *Social Indicators in Korea* (1988), 170–1.

37. Comrade Kim Il-Sung's speech, 'The Socialist System of Our Country', which is fully quoted in an article by Yong-Pok Chon in *Rodong Sinmun* (6 March 1990), 2. The words in brackets are my own insertion.

38. Ibid.

39. According to Western sources, some very limited retail goods in some secondary markets sell at a price 50 to 100 times higher than the official market price in dollar terms. Take rice, for instance: the distribution price per kilogram is 8 *chon* (1 *chon* = 0.01 *won*), but its secondary market price is W25.

40. Fidel Castro, Opening Address to the Second Congress of the Association of Third World Economists, Havana, 26 April 1981, 16; quoted from Halliday, 'Economics', 33.

41. The Gini index is computed for incomes by means of the following formula:

$$G = \frac{2}{(T-1)} \sum_{t=1}^{T} t y_t - \frac{(T+1)}{(T-1)},$$

where y_t is the proportional share in income of the *t*th family. Note that the measure depends on how observations are measured; e.g., the estimated coefficient will vary according to whether the data are adjusted for family size, or for differences in pre-tax and post-tax distributions, or for variation in non-earned services provided to individuals by the state, etc.

42. The only available data in both time-series (1979–87) and cross-sectional (income group) series in South Korea is from the *Annual Report on the Family Income and Expenditure Survey*, annually published by the Statistical Bureau of the Economic Planning Board, Seoul. It must be noted that Gini index estimates in Table 3.19 are based on pre-tax household incomes, which are also not adjusted for the number of family members and ages of family members.

43. See Nanak C. Kakwani, *Income Inequality and Poverty: Methods of Estimation and Policy Applications* (Oxford University Press for World Bank, 1980), 386–9 for income shares, Gini index, and a measure of skewness of the Lorenz curve in fifty countries, and for inter-country comparison of income inequality.

4

Public Finance in North and South Korea

Lead out those who have eyes but are blind,
who have ears but are deaf.
All the nations gather together
and the people assemble.
Which of them foretold this
and proclaimed to us the former thing?
Let them bring in their witnesses
to prove they were right.
So that others may hear and say 'it is true'.

Isa. 43: 8, 9

4.1 INTRODUCTION

Public finance affects a nation's use of aggregate resources and
its financing patterns; together with monetary and exchange rate
policies, it influences the external balance of payments, the accumulation of foreign debt, and the rates of inflation, interest, and
exchange. However, the spread of impact might differ significantly
depending upon whether the economy is market-oriented or centrally planned. Public spending, taxes, user charges, and borrow-
ing can affect not only the macroeconomic behaviour of producers
and consumers, but also the distribution of wealth and income in
an economy. Of course, the channels of policy influences are more
indirect in a market-oriented economy than in any controlled
economy, which is often bound by economic ideologies that set
objectives or priorities.

In this chapter, the structural aspects of North and South
Korean public finance policy—how and where spending is allocated
and how revenue is raised—will be discussed and compared. The
differing orientations of their economic philosophies have without
doubt resulted in two different ways of life, thought, and value

systems among the separated peoples who once shared the same cultural and social background. Both sides are set in explosive competition with one another to achieve rapid economic growth and military power via different approaches. The race demonstrates the complexity of the relationship between North and South. Whether a series of recent inter-Korean contacts will turn out to be a turning point in the relationship is yet to be seen.

Since the role of the state budget in the economic and military competition is so great, it is necessary first to examine how the budgetary spending is allocated in the two rival states. In so far as the national budget serves to allocate national resources, it should be possible also to devise some criterion by which to judge the economic well-being of the people from the allocational structure of their national budgets.

4.2   NORTH KOREAN DEVELOPMENT

Our brains just could not handle this idea of a market.
Mikhail Gorbachev, 11 September 1990, Moscow

North Korea's public finance policy started effectively with its early land reform and centralization of all production and service sectors, including mines, railroad and communication networks, and banking, in its initial start-up stage. In order to rehabilitate and reconstruct the post-war economy in line with Soviet-type central planning, central fiscal policy has been used exclusively as a strong state policy tool. The North Korean fiscal policy seems mostly to have targeted the support of strategic industries as well as programmes that could promote the material well-being, education, and physical and mental health of the masses.[1]

One of the functions of North Korean public finance is to allocate and guide the budget for public agencies. The other function is to supervise the financing of state and social enterprises, credit banking, insurance, education, and production as well as the distribution of net material product. The role of centralized public finance in the framework of economic policy and planning is pervasive under a government dictatorship. In a command economy, income is distributed through the channel of state budget spending. Therefore, the budgetary mechanism plays a very critical role in economic development as well as in maintaining a necessary trade-off between accumulation and distribution of national wealth.

4.2.1   State Revenue and Expenditure

The state budget proposal is extensively discussed and eventually passed as a budgetary order by the DPRK Supreme People's Assembly at its regular session, usually held every April. The budget year starts on 1 January and ends on 31 December each year.

By 'budget' is meant the government's basic plan. This plan establishes the financial demands from various sectors of the economy and allocates funds to each sector. In North Korea, the government budget reflects virtually all sectors of economic activity. The Ministry of Finance, under the Office of State Affairs, has sole responsibility for drawing up a budget plan, presenting it for discussion, and getting it passed. It is, in fact, in charge of every aspect of finance as a whole. By 1972, North Korea was maintaining a consolidated budget system in which the central government budget encompassed every local government's budget.[2] However, in 1973 local government budgets were divorced from the central one so as to give local agencies autonomous room to secure their budgets to meet expenditure on the local level, thus achieving self-sufficiency and autonomy in both production and consumption at the provincial level. Local agencies have to pass on any budget surplus to the central government, which in turn enables the government to recoup part of its overall expenses.

The state revenue comes mainly from *socialist accounting*, i.e. accounting and recording of all production activities in a state socialist economy like North Korea, turnover profits ('transaction revenues'), resident taxes, and foreign remittances, of which accounting revenue accounts for the largest part.[3] State expenditure consists of four major categories: economic development, social and cultural development, defence, and administration. Out of total spending, about 60% is known to go to investment in economic development, including industrial, agricultural, transportation, communication, and construction sectors.

North Korea has traditionally focused its investment on heavy industry, which is of course associated with defence. DPRK official statistics about budgetary spending reveal that the share allocated to defence has steadily decreased over time, from a peak of 32.4%

in 1968 to 12.0% in 1989 (see Table 4.1). Since 1972, the North's military sector has accounted for no more than 15% of total annual expenditure. This, interestingly, contrasts sharply with the South Korean estimate of the North's military expenditure, as more than 20% of GNP. Furthermore, some economists in the South suspect that Pyongyang is concealing a considerable proportion of actual military expenditure in other categories such as economic development. This might be true if heavy industry in the North were regarded as a part of the defence sector, as already pointed out in the previous chapter. Yet, the distinction is quite arbitrary.

Table 4.1 shows the trends of North Korea's revenue and expenditure during the period 1948–92. From 1953 to 1989, North Korean government expenditure increased more than 67-fold. Government expenditure as a share of its GNP was 52.3% during the 1960–9 period, 49.3% in 1970–9, and 56.5% in 1980–9. A striking characteristic of the North Korean budget is the increasing share of both revenue and expenditure in GNP and the changing composition of budgetary spending in the 1980s. In 1980, the central government revenue and expenditure as percentages of North Korea's GNP were 51.0% and 52.9% respectively: they rose to 62.0% and 61.6% in 1989. These figures can be compared with the percentages of government expenditure in GNP (or GDP) in other communist economies, as given in Table 4.2 for the 1980s. The weight of government expenditure in the North Korean economy is much more similar to the cases of Hungary and the Soviet Union than to other communist countries, although the share of government spending in the Soviet Union might be slightly larger than the figures given in Table 4.2 if GNP data instead of GDP data were used as the denominator.

It is noteworthy that the share of expenditure on economic development has grown 7.9-fold, while social welfare expenditure has increased 5.3-fold, military spending 2.2-fold, and administration costs 5.3-fold during the period 1970–89. The military spending will be further discussed in comparison with South Korea's budget expenditure in Section 4.4 below.

4.2.2 *Income Elasticities of Government Spending and Revenue, 1949–1989*

To explore the extent of correlation between the stream of government budget and GNP on a per capita basis, per capita income

TABLE 4.1(a) North Korea's state revenue and expenditure, 1948–1993 (thousands of North Korean *won*)

	Revenue	Expenditure	Budget outlays by function			
			Economic development	Social and cultural policy	National defence	General administration
1948	155 710	136 540	48 880	36 190	9 010	42 460
1949	203 010	196 570	83 080	37 760	11 400	31 050
1953	527 270	495 970	239 900	55 970	75 240	47 930
1955	1 081 570	1 006 190	753 450	85 100	61 900	64 510
1956	992 540	955 980	710 470	121 460	56 500	58 550
1957	1 251 160	1 022 450	720 970	177 210	62 340	61 930
1958	1 529 140	1 321 310	932 300	246 840	64 690	77 480
1959	1 715 700	1 648 990	1 135 300	382 340	62 090	69 260
1960	2 019 310	1 967 870	1 361 570	482 330	61 000	62 970
1961	2 400 000	2 338 000	1 707 900	500 330	58 450	71 310
1962	2 896 360	2 728 760	1 976 040	616 900	136 820[a]	
1963	3 144 820	3 028 210	2 240 880	645 010	142 320[a]	
1964	3 498 780	3 418 240	2 377 210	696 610	344 420[a]	
1965	3 573 840	3 476 130	2 363 760	684 220	278 090	150 060
1966	3 720 210	3 556 960	n.a.	n.a.	n.a.	n.a.
1967	4 106 630	3 948 230	n.a.	n.a.	1 200 260	n.a.
1968	5 023 700	4 828 640	2 353 500	829 130	1 559 380	86 630
1969	5 319 030	5 048 570	n.a.	n.a.	1 565 060	n.a.

Year						
1971	6 357 350	6 301 680	2 783 790	1 444 670	1 959 820	113 400
1972	7 430 300	7 388 610	4 099 320	1 878 070	1 256 060	155 160
1973	8 599 310	8 313 910	4 755 540	2 103 440	1 280 340	174 590
1974	10 015 250	9 672 190	5 513 930	2 397 920	1 557 220	203 120
1975	11 586 300	11 367 480	6 506 880	2 257 610	1 864 270	238 720
1976	12 625 830	12 325 500	6 918 280	3 083 010	2 065 370	258 840
1977	13 789 000	13 449 170	7 681 710	3 391 310	2 095 820	280 330
1978	15 657 300	14 743 600	8 450 870	3 638 880	2 344 230	309 620
1979	17 301 320	18 106 440	10 921 270	4 170 160	2 629 800	385 210
1980	18 139 230	18 836 910	11 191 590	4 411 060	2 750 190	484 070
1981	20 684 000	20 333 000	12 232 410	4 711 830	3 009 280	475 480
1982	22 680 000	22 211 600	13 639 140	4 894 460	3 252 880	425 120
1983	24 383 600	24 018 600	15 155 740	4 851 750	3 530 730	480 370
1984	26 551 000	26 158 000	16 845 750	5 205 440	3 819 060	287 740
1985	27 438 870	27 328 830	17 080 520	5 657 070	3 935 350	655 890
1986	28 538 500	28 396 100	18 031 520	5 764 410	4 003 850	596 320
1987	30 337 200	30 085 100	19 976 510	5 686 080	3 971 230	451 280
1988	31 905 800	31 660 900	21 244 460	6 015 570	3 862 630	538 240
1989	33 608 100	33 382 940	22 501 840	6 307 740	4 005 950	567 410
1990	35 656 100	35 656 100	24 058 970	6 718 860	4 314 380	563 890
1991	37 194 840	36 909 240	25 047 070	6 927 040	4 466 020	469 110
1992[b]	39 592 000	39 592 000	26 675 130	7 735 800	4 582 100	513 110
1993[b]	40 449 850	40 449 850	27 423 790	7 756 470	4 692 180	582 410

[a] Includes national defence and general administration.
[b] Planned budget announced by the KCBS on 8 May 1992 and 8 April 1993.

Source: North Korea's *Chungang Nyungam* (Central Year Book), each issue. For 1989 and 1990 data, see FBIS-EAS-90-102, 25 May 1990, 26, which translated into English the summing up of the fulfilment of the state budget of North Korea for 1989 and the state budget for 1990, announced at the second-day sitting of the first session of the Ninth Supreme People's Assembly, Pyongyang.

TABLE 4.1(b) North Korea's budget outlays by function, 1948–1993 (%)

	Total expenditure	Economic development	Social welfare and culture	National defence	General administration
1948	100.0	35.8	26.5	6.6	31.1
1949	100.0	42.2	19.2	5.8	15.8
1953	100.0	48.4	11.3	15.2	9.6
1955	100.0	74.9	8.5	6.2	6.4
1956	100.0	74.3	12.7	5.9	6.1
1957	100.0	70.5	17.3	6.1	6.1
1958	100.0	70.6	18.7	4.9	5.8
1959	100.0	68.8	23.2	3.8	4.2
1960	100.0	69.2	24.5	3.1	3.2
1961	100.0	73.0	21.4	2.5	3.1
1962	100.0	72.4	22.6	5.0[a]	
1963	100.0	74.0	21.3	4.7[a]	
1964	100.0	69.5	20.4	10.1[a]	
1965	100.0	68.0	19.7	8.0	4.3
1966	100.0	n.a.	n.a.	n.a.	n.a.
1967	100.0	n.a.	n.a.	30.4	n.a.
1968	100.0	48.7	17.2	32.3	1.8
1969	100.0	n.a.	n.a.	31.0	n.a.
1970	100.0	47.3	19.9	31.0	1.8
1971	100.0	44.2	22.9	31.1	1.8

Year					
1972	100.0	55.5	25.4	17.0	2.1
1973	100.0	57.2	25.3	15.4	2.1
1974	100.0	57.0	24.8	16.1	2.1
1975	100.0	57.1	19.9	16.4	2.1
1976	100.0	56.1	25.0	16.8	2.1
1977	100.0	57.1	25.2	15.6	2.1
1978	100.0	57.3	24.7	15.9	2.1
1979	100.0	60.3	23.1	14.5	2.1
1980	100.0	59.4	23.4	14.6	2.6
1981	100.0	60.2	23.2	14.8	2.3
1982	100.0	61.4	22.1	14.6	1.9
1983	100.0	63.1	20.2	14.7	2.0
1984	100.0	64.4	19.9	14.6	1.1
1985	100.0	62.5	20.7	14.4	2.4
1986	100.0	63.5	20.3	14.1	2.1
1987	100.0	66.4	18.9	13.2	1.5
1988	100.0	67.1	19.0	12.2	1.7
1989	100.0	67.4	18.9	12.0	1.7
1990	100.0	67.5	18.8	12.1	1.6
1991	100.0	67.9	18.8	12.1	1.2
1992	100.0	67.5	19.6	11.6	1.3
1993	100.0	67.8	19.2	11.6	1.4

a Includes national defence and general administration.

Source: as for Table 4.1(a).

TABLE 4.2 Government expenditure as a percentage of GNP in communist countries, 1980–1989

	North Korea	Soviet Union[a]	Poland	Yugoslavia	Hungary	Romania	Czechoslovakia	Bulgaria
1980	52.9	47.6	—	8.7	—	—	—	23.0
1981	55.7	47.6	—	8.1	64.2	—	—	26.1
1982	54.3	49.3	—	7.4	60.4	—	35.2	26.0
1983	52.3	48.6	39.2	7.6	62.7	14.9	35.7	27.9
1984	55.5	48.8	40.0	7.3	60.7	14.3	36.9	28.2
1985	56.4	49.7	38.7	6.7	57.9	15.7	38.3	30.4
1986	57.4	52.2	32.9	6.7	69.4	14.8	38.4	31.2
1987	58.9	52.2	29.7	6.6	67.4	17.9	39.5	32.8
1988	60.2	52.5	28.0	—	—	—	40.5	—
1989	61.6	52.8	—	—	—	—	—	—

[a] Government expenditure as percentage of GDP. See *PlanEcon Report* (21 February 1990) (Washington, DC), 3; see also Gur Ofer, 'Budget Deficit, Market Disequilibrium and Soviet Economic Reforms', *Soviet Economy*, 5 (April–June 1989), 107–26.

Sources: World Bank, *World Tables, 1988–89*, for GNP data; John Paxton (ed.), *The Statesman's Year-Book, 1990–1991* (New York: St Martin's Press, 1990), for government expenditures; *PlanEcon Report: Review and Outlook* (Fall 1989); US CIA, *Handbook of Economic Statistics* (1989).

elasticities of both revenue and expenditure by function were estimated using North Korea's budget data in natural logarithms from Table 4.1. The results are as follows:

(1) $REVENUE_t = 3.8090 + 0.9874GNP_t$ $RMSE = 0.79384$
      (25.059) (10.719) R^2 $= 0.8097$

(2) $EXPENDITURE_t = 3.7790 + 0.9934GNP_t$ $RMSE = 0.84271$
      (23.42)   (10.158) R^2 $= 0.7926$

(3) $ADMIN_t = 0.12082 + 0.54674GNP_t$ $RMSE = 1.21449$
      (0.52)   (3.88) R^2 $= 0.3579$

(4) $DEFENCE_t = 1.7488 + 1.19652GNP_t$ $RMSE = 1.20699$
      (7.567)   (8.543) R^2 $= 0.7300$

(5) $WELFARE_t = 2.2178 + 1.1933GNP_t$ $RMSE = 1.0946$
      (10.152) (9.395) R^2 $= 0.7658$

(6) $ECONOMY_t = 3.1315 + 1.1110GNP_t$ $RMSE = 1.25292$
      (13.053) (7.641) R^2 $= 0.6838$

Numbers in parentheses are t-statistics of the estimated values. The results can be summarized as follows.

Item	Elasticities (per capita)
Revenue	0.9874
Expenditure	0.9934
Spending for administration	0.5467
Spending for national defence	1.1965
Spending for welfare	1.1933
Spending for economy	1.1110

It is interesting to find that the response of per capita central government spending with respect to per capita GNP is near unitary elasticity with the exception of general administration expenditure in North Korea. As will be explained when discussing South Korea below, North Korea has slightly higher estimates in military, social welfare, and economic development expenditure than its competitors in the South, while South Korea beats the North in government revenue and general administration expenditure elasticities (see next section).

4.2.3   Financial Functions

Economic decisions on both macro and micro questions of resource allocation are in the hands of the state authorities. In a
centralized economy the central plan specifies most value aggregates and physical inputs and outputs in the economy. Money
plays only a passive accounting role. The structure of the plan is
strictly hierarchical, so that the lower levels are formally subordinated to those above. Decision-making is concentrated near the
top. The plans are enforced by rationing the means of production
through administrative orders, and capital through the allocation
of investment funds.[4] Since capital is centrally allocated in North
Korea, investment funds are generally provided in the form of
grants through government-controlled financial agencies. Therefore, the financial system in North Korea provides government
services that facilitate the allocation of investment funds. Financial intermediaries in a centrally planned economy are very different from those functions usually expected of them in most
market-oriented economies. In capitalist countries, mutually
competitive financial intermediaries provide a variety of services
and reduce the costs of transactions. As a profit-maker, a financial
intermediary has to evaluate alternative investments, to monitor
the activities of borrowers, and to increase the efficiency of resource
use. Such routine business practices are lacking in the centrally
planned financial system.

Strong regulation has so far limited the structure of financial
intermediaries to a few public banks, including the Chosen Central Bank, the Foreign Trade Bank, Reconstruction Bank, Daesung
(commercial) Bank, and Kumkang (commercial) Bank, as introduced in Chapter 2. Banking business services include financing
investment funds and extending trade credit to enterprises; demand and savings deposits; and remittance and foreign exchange
dealings. Since all banking business is under tight government
control, there seems little incentive for financial innovation in the
planned economy.

In North Korea, basic wages and salaries are regulated by fixed
wage scales established for different categories of skill and effort.
Prices for other commodities and services are set administratively.
The foreign exchange rate is fixed centrally and plays no significant role in domestic price formation. Thus, the central bank has

nothing to do with price stabilization, although it is jurisdictionally in charge of money supply for the economy. Because of the heavy reliance on administrative control, the role of money is strictly limited to that of an accounting unit[5] while administrative prices do not play any allocative role. None the less, a prominent characteristic of North Korea's prices has been their relative stability, and no significant changes have occurred since the 1950s. This proves that the usual mechanism of two-way links between the spheres of production and the circulation of currency does not exist in the planned economy. In other words, prices do not reflect the real relationship between supply and demand: instead, they merely reflect the directives formulated in the depths of the planning offices. Output is determined not by the relationship between prices and costs, but by the complex interaction of natural and arbitrary subjective factors of top decision-makers. In such an economy, where production and money circulation work independently without any discernible link with natural economic proportions, the role of public finance ceases to be that of a universal equivalent, and instead performs functions somewhat different from those carried out in a market economy. As long as the North Korean economy remains primarily a non-monetary, rationing economy, its financial system will play an extremely limited role. But times are now changing, and institutional transformation in the real and financial sectors may be badly needed to enable North Korea to adapt to international environments if the country wants to keep its economy operational in the open world.

4.3 SOUTH KOREAN DEVELOPMENT

> In real life, budget decisions are undoubtedly influenced to a greater or lesser extent by such non-economic and nonrational factors as pride and prejudice, provincialism and politics.
>
> *Quoted from Lance T. LeLoup, 'Budgetary Politics: Dollars, Deficits, Decisions' 1977*

In the 37 years from 1953 to 1989, South Korea's gross national product increased more than fifteenfold. Having adjusted successfully both to the oil shocks of the 1970s and to the debt shock of the early 1980s, it has emerged as one of the top developing countries, just

behind Singapore and Taiwan, although it faces many new challeng-
ing domestic issues in the process of broader democratization and
liberalization, especially since the mid-1980s.

Despite its real strength and economic robustness, this rapidly
developing country, densely populated, and lacking vital natural
resources, has spawned a heap of agonies, not least of which is
the conflict between the haves and have-nots of the population.
Internationally, South Korea's economic success inevitably has invited strong criticism from other countries, particularly the United
States. This is because South Korea has not yet begun to observe
the rules of free trade. South Korea's export-driven development
strategy does not represent a *laissez-faire* approach. The credit
system has channelled financial resources at subsidized rates to
preferred activities. The government has extended to firms various
preferential duties for export promotion, has offered favourable
tax rates on profits and income earned from exports, and has
provided direct cash subsidies for those export-oriented firms.[6] In
this way, public finance has played an extraordinary role in making
South Korea one of the little dragons in Pacific Asia. In this section,
I shall first briefly review the financial system and its relative
importance in the South's market economy, with due focus on the
structure of government revenue and expenditure. Second, I shall
analyse the empirical impact of government finance and spending
on the economy. Third, I shall give a brief introduction to the
South's financial market, including the banking system.

4.3.1   The Structure of Public Finance

The public sector in South Korea consists mainly of government
and public enterprise sectors. The government sector is divided
into the central government and local government sectors in terms
of policy application. The central government sector encompasses
the general account, special accounts, and government management
funds, while there are also general and special accounts for the
local governments, including the Special City of Seoul, four other
major cities, and nine provinces, along with financial accounts for
counties and cities under each provincial government in the local
financial sector.

Budget allocations are made to the public enterprise sector

through special accounts available for the state-run enterprises: the railways, the monopoly (i.e. tobacco) office, telecommunication services, government purchase office, and grain management funds.

Finances for the central government and local governments at all levels are systematically linked. First, a local administrative body collects local taxes within the scope of the taxation law, which stipulates all details related to such finances, such as the levying of local taxes, the liability of taxpayers to local taxes, tax rates, and the standard assessment of taxation. Second, local bonds, when floated to supplement local finances out of the official budget, are issued only with the authorization of the Ministry of Finance regarding the amount of the issuance and uses of the money. Third, the central government at present provides subsidies and grants to local governments in statutory percentages of the internal taxes; it also finances local governments in the form of statutory subsidies for education and local education grants.

Next, I shall briefly describe the composition and trends of government revenue and expenditure in South Korea. As the nation's economy expands, the level of the public financial burden increases. The ratio of total finances (budget of the general government, consisting of central and local governments) to current GNP has grown from 15.7% in 1958 to 30.8% in 1969 and to 27.8% in 1989 (see Table 4.3(*a*)). The central government budget share (general accounts plus special accounts) of the total budget of general government (central government plus local governments) accounted for 92.8% in 1958, 85.9% in 1970, and 78.8% in 1989. The local government budget share was 7.2% in 1958, 14.1% in 1970, and 21.2% in 1989. The composition of the general accounts of the central government budget amounted to 71.5% in 1970, 75.0% in 1980, and 86.3% in 1989, leaving a residual portion for special accounts in total central government finance.

As is well known, the traditional illustration of resource scarcity is the social choice of 'guns' or 'butter'. Recent debate over budget priorities in South Korea often centres on the relative shares to economic development and social welfare programmes and national defence. How does the South Korean government spend its money—on butter or guns? Table 4.3(*b*) compares government spending by function over the past three decades.

In the past 30 or more years, expenditure on national defence has constantly occupied a significant proportion of the South

TABLE 4.3(*a*) Ratios of general government budget to GNP, South Korea, 1969–1990

	Budget of general government	Central government			Local government
		Total	General accounts	Special accounts	
1969	30.8	26.9	16.8	10.1	3.9
1970	25.1	21.6	15.4	6.2	3.5
1971	25.3	21.6	15.1	6.4	3.7
1972	26.6	23.1	16.7	6.5	3.5
1973	21.4	17.9	12.2	5.7	3.5
1974	22.6	18.9	13.4	5.5	3.7
1975	24.6	21.1	15.3	5.8	3.5
1976	24.3	21.0	15.5	5.4	3.4
1977	25.0	21.0	15.5	5.5	4.0
1978	25.3	19.9	14.8	5.1	5.4
1979	26.8	21.0	16.4	4.6	5.9
1980	29.0	23.5	17.6	5.9	5.6
1981	29.2	23.5	17.4	6.1	5.8
1982	28.7	22.1	17.6	4.5	6.2
1983	27.3	20.4	16.5	3.9	6.8
1984	27.9	20.4	15.8	4.6	7.1
1985	26.2	19.2	15.9	3.3	6.9
1986	25.9	19.0	15.2	3.8	6.6
1987	25.0	18.1	14.9	3.2	6.8
1988	23.2	18.0	14.3	3.7	5.2
1989	27.8	19.7	15.6	4.0	8.1
1990[a]	27.9	18.6	14.5	4.2	9.2

[a] Preliminary or budget (plan) data.

Korean budget. It competed with expenditure on economic development during the 1960s and 1970s, but in the 1980s it gradually took the lion's share of budget expenditure. A sector that has slowly but steadily increased its marginal share of the budget is education. So far, social welfare has been given a low priority. All in all, there has not been any major shift in the overall composition of central government expenditure over the past three decades, as shown in Table 4.3(*b*).

The relatively low proportion of spending on health and social welfare indicates that the government has neglected these areas in favour of promoting efficiency and growth in economic development and national defence. The natural outcome of such disequilibrium development strategies is revealed in today's sharp social conflict between the haves and have-nots in South Korea.

The proportions of general accounts expenditure to GNP were 25.3% in 1962, 15.9% in 1970, 15.3% in 1975, 17.6% in 1980, 15.9% in 1985, and 15.3% in 1989. The characteristics of South Korea's budget spending are obvious. It is common knowledge that budget spending rises as incomes rise in most countries; but the share of the South Korean budget in GNP has remained at a relatively very low level even when compared with those of far less developed (low-income) countries.[7] Table 4.4 gives a cross-comparison made in 1986.

The role of central government in South Korean economic development has never been small in spite of the relatively small percentage of its budget spending to GNP. Under the limited capacity of the budget, the government could not help allocating a lower priority to social welfare than to other sectors.

On the other hand, South Korea has spent a very high proportion of its budget on national defence, which must have inevitably constrained spending for other sectors.

4.3.2 *Income Elasticities of the Government Budget, 1958–1989*[8]

In order to examine the structure of the South Korean budget, income elasticities (per capita) were estimated for the 1958–89 period. These results can be used to make comparisons with similar estimates for North Korea, which were introduced in the previous section.

TABLE 4.3(b) The central government budget of South Korea, 1961–1990

| | | Billions of South Korean *won* | | Composition of outlays by function[d] (%) | | | | | |
| | Tax revenue[a] | Expenditure | | | | | | | |
		General accounts[b]	Consolidated[c]	General administration	National defence	Economic development	Social welfare	Education	Other
1957	10.5	20.5	—	20.9	44.1	—	—	—	—
1958	14.1	27.1	41.0	9.8	32.3	33.5	10.2	10.9	3.3
1959	22.2	30.5	41.6	12.9	35.8	23.8	9.2	15.6	2.0
1960	24.1	33.6	42.7	15.2	34.1	22.3	4.0	21.5	3.0
1961	24.6	57.7	59.0	12.7	29.2	32.8	9.6	13.6	2.1
1962	28.3	88.4	76.0	11.5	27.6	34.7	10.5	13.9	1.8
1963	31.1	72.8	79.9	13.4	26.3	34.2	11.0	13.8	1.2
1964	37.4	75.2	82.0	12.4	31.2	28.9	11.0	15.2	1.3
1965	54.6	93.5	100.9	13.2	30.6	29.9	9.8	15.4	1.1
1966	87.7	140.9	163.5	12.2	26.2	33.9	10.4	16.3	1.0
1967	129.3	180.9	250.7	12.5	25.6	33.5	11.2	16.3	0.8
1968	194.3	262.1	296.7	11.9	23.4	32.3	15.7	16.0	0.8
1969	262.8	370.5	470.7	12.2	21.4	29.7	20.8	15.0	0.9
1970	334.7	441.3	522.0	13.4	23.6	28.6	15.7	17.3	1.5
1971	407.7	546.3	640.1	12.8	25.2	29.4	12.6	18.5	1.4
1972	433.4	701.1	851.5	22.5	27.0	23.4	7.6	17.1	2.4
1973	521.5	655.4	854.3	22.8	28.8	19.5	8.4	17.5	3.0

1974	844.7	1 013.9	1 356.7	18.5	27.6	31.1	6.8	13.4	2.6
1975	1 225.5	1 535.3	2 054.3	20.0	27.1	30.3	7.3	13.1	2.2
1976	1 914.7	2 142.2	2 663.5	15.8	31.4	28.4	6.0	14.4	4.0
1977	2 402.7	2 739.9	2 737.6	16.2	31.7	27.4	6.1	14.5	4.1
1978	3 372.3	3 538.7	3 818.6	9.5	32.6	26.2	7.9	13.7	10.0
1979	4 401.7	5 053.7	5 224.0	9.1	26.7	31.8	8.1	14.4	9.9
1980	5 297.7	6 486.1	6 562.0	8.5	30.6	26.0	9.9	14.6	10.4
1981	6 577.9	7 909.8	8 044.8	8.9	28.0	24.7	13.9	14.4	10.1
1982	7 636.4	9 178.9	10 115.0	9.2	27.3	21.6	13.7	17.0	11.1
1983	9 220.7	10 180.8	10 681.2	10.1	27.9	19.9	11.8	17.9	12.3
1984	10 053.7	11 072.1	11 874.6	9.0	26.6	19.1	15.0	16.8	13.6
1985	11 047.4	12 406.4	13 336.9	9.4	26.6	21.9	12.4	16.6	13.2
1986	12 622.3	13 796.5	14 948.8	10.0	27.5	18.1	12.5	17.0	14.9
1987	15 439.4	15 794.5	16 943.6	8.9	27.0	20.0	13.1	16.7	14.5
1988	18 537.1	18 025.0	19 453.7	9.7	27.1	20.8	15.2	19.0	15.6
1989	25 590.9	21 653.1	22 279.2	11.2	32.4	15.9	8.8	21.1	10.6
1990[e]	23 025.4	23 025.4	—	11.2	30.2	15.2	10.3	22.1	11.1

[a] Tax (internal taxes + customs duties) revenue only, which does not include monopoly profits, non-tax revenues, trust funds, etc.

[b] Central government general accounts expenditure, including for general administration, national defence, economic and social development, unallocable grants to local government, repayment of debt, etc. Note that special accounts are not included in the amounts here.

[c] Central government consolidated expenditure consists of current expenditure and capital expenditure, but excludes net lending. These data were used to estimate the income elasticities of expenditures given in the text.

[d] The composition is calculated from the data on general accounts by function. The data for 1957–71 are based on the old classification series, and data for 1972–90 are based on the new classification series.

[e] Preliminary and plan data.

Sources: Korea Development Institute, *National Budget and Policy Goal, 1989*; Bank of Korea, *Economic Statistics Yearbook* (each issue); Ministry of Finance, *Finance Statistics* (each issue); Economic Planning Board, *Major Statistics of Korean Economy* (each issue).

TABLE 4.4 Budget spending by different income economies, 1986[a]

	Low-income countries	Middle-income countries	Upper-Middle income countries	Industrial market economies	South Korea[b]
Per capita GNP ($)	200	750	1890	12 960	2505
Spending GNP (%)	20.8	24.9	28.3	28.6	15.2
National defence (%)	17.7	15.8	10.3	16.4	27.5
Education (%)	9.8	14.5	10.2	4.5	17.0
Health (%)	3.6	4.0	5.1	12.9	2.0
Housing and social welfare (%)	6.2	9.1	17.3	39.0	10.5
Economy and service (%)	23.8	21.5	19.6	9.5	18.1

[a] Note that South Korea's government spending accounts for general accounts only. If special accounts are added to it, the percentage of central government expenditure to GNP is 19.0%. If, in addition, local government expenditure is added, the percentage of general government budget to GNP rises to 26.2% in 1986.

[b] General accounts only.

Sources: World Bank, *World Development Report;* but for South Korea figures are calculated from Table 4.3 above.

The per capita variables used are all in natural logarithms:

(1) $REVENUE_t = -4.0417 + 1.0140GNP_t$ $RMSE = 0.17513$
      (-124.343) (63.933) $R^2 = 0.9927$

(2) $EXPENDITURE_t = -4.0555 + 1.0232GNP_t$ $RMSE = 0.16899$
      (-129.301) (66.86) $R^2 = 0.9933$

(3) $ADMIN_t = -5.9569 + 0.8707GNP_t$ $RMSE = 0.14426$
      (-222.478) (66.646) $R^2 = 0.9933$

(4) $DEFENCE_t = -5.3730 + 1.0146GNP_t$ $RMSE = 0.20527$
      (-141.023) (54.574) $R^2 = 0.9900$

(5) $WELFARE_t = -7.0532 + 1.0813GNP_t$ $RMSE = 0.20995$
      (-180.997) (56.87) $R^2 = 0.9908$

(6) $ECONOMY_t = -5.3528 + 0.9557GNP_t$ $RMSE = 0.25281$
      (-114.079) (41.741) $R^2 = 0.9831$

As usual, numbers in parentheses are t-values. The results can be summarized as follows.

Items	Income elasticities
Revenue	1.0140
Expenditure	1.0232
Spending for general administration	0.8707
Spending for national defence	1.0146
Spending for welfare	1.0813
Spending for economic development	0.9557

The elasticities of per capita government outlays by function with respect to per capita GNP are positive, with the exception of general administration and economic development spending; of all of them, the elasticity on welfare outlay is the highest. This implies that South Korea has made an effort to raise expenditure on social welfare as income has grown. Thus, it has tried to induce structural changes to improve the role of social welfare in finance, although the share of social welfare expenditure in GNP was extremely low in absolute quantity from the beginning and the ratio of budget to GNP has remained relatively unchanged over time.

4.3.3 The Effect of Government Spending on the Economy

Studies on the effect of public finance on the aggregate economy, directly or indirectly, abound in macroeconomics literature. Also, public finance policy is closely related to the 'big trade-off' between efficiency and equality which became an important part of the economics lexicon with the publication of Arthur Okun's book, *Equality and Efficiency*, in 1975.[9]

The first question has been widely pursued in the well-known controversy between the Keynesian multiplier approach and the monetarist assertion that fiscal expansion policy only ends up with 'crowding-out' effects.[10] All positive and normative aspects of the effects of government taxation and expenditure on an economy constitute an important topic of macroeconomic research. Welfare effects of taxation as well as alternative forms of public spending need to be performed in an overlapping-generations model framework. But these are research topics beyond the scope of this study. Here, we will explore a simple relationship between government fiscal policy and the aggregate economy in South Korea.

As a simple illustration, let us assume that the goods clear the market and that labour is in excess supply in a monetary economy with two markets (goods and labour) and three agents (a firm, a household, and a government). In this world, the firm can realize its 'neoclassical' plan for employment and sales of goods. Therefore,

$$N = F'\left(\frac{W}{P}\right)$$

$$Y = \Phi\left[F'\left(\frac{W}{P}\right)\right].$$

We may further assume the demand side of the economy as $Y = C(Y, P, T) + G$.

Combining the two supply-side equations with the demand-side equation of the goods market, we obtain the following system for the equilibrium values, N^*, Y^*, and P^*:

$$Y^* = C(Y^*, P^*, T) + G$$

$$Y^* = \Phi\left[F'\left(\frac{W}{P^*}\right)\right]$$

$$N^* = F'\left(\frac{W}{P^*}\right),$$

where
Y   = demand
C   = consumption expenditure
G   = government expenditure
P   = price level
T   = tax
W/P = real-wage level.

It is easy to see that the minimum price $\bar{P}$ no longer has any effect, a predictable result because the goods market has cleared. To calculate the multipliers associated with alternative policies, we may simply define

$$Y_s(P, W) = \Phi\left[F'\left(\frac{W}{P}\right)\right]; \quad \frac{\partial Y_s}{\partial P} > 0, \frac{\partial Y_s}{\partial W} < 0.$$

The multipliers corresponding to Keynesian policies are

$$\frac{\partial Y^*}{\partial G} = \frac{1}{1 - \dfrac{\partial C}{\partial Y} - \dfrac{\partial C/\partial Y}{\partial Y_s/\partial P}} > 0$$

$$\frac{\partial Y^*}{\partial T} = \frac{\partial C/\partial T}{1 - \dfrac{\partial C}{\partial Y} - \dfrac{\partial C/\partial P}{\partial Y_s/\partial P}} < 0.$$

We can also see that, by increasing demand, these Keynesian policies induce some price increases, which reduce consumption and thus the multiplier effects:

$$\frac{\partial P^*}{\partial G} = \frac{1}{\dfrac{\partial Y_s}{\partial P}\left(1 - \dfrac{\partial C}{\partial Y}\right) - \dfrac{\partial C}{\partial P}} > 0$$

$$\frac{\partial P^*}{\partial T} = \frac{\partial C/\partial T}{\dfrac{\partial Y_s}{\partial P}\left(1 - \dfrac{\partial C}{\partial Y}\right) - \dfrac{\partial C}{\partial P}} < 0.$$

Note that these price effects may be sufficient to diminish private consumption in case of an increase in public spending.

$$\frac{\partial C^*}{\partial G} = \frac{\partial C/\partial Y + \dfrac{\partial C/\partial P}{\partial Y_s/\partial P}}{1 - \dfrac{\partial C}{\partial Y} - \dfrac{\partial C/\partial P}{\partial Y_s/\partial P}}.$$

The denominator of this expression is always positive, but the numerator may be either positive or negative. In this last case there is a partial crowding-out effect.

Meanwhile, the classical or monetarist measures of reducing nominal wages become effective in this regime:

$$\frac{\partial Y^*}{\partial W} = \frac{-(\partial Y_s/\partial W)(\partial C/\partial P)}{\dfrac{\partial Y_s}{\partial P}\left(1 - \dfrac{\partial C}{\partial Y}\right) - \dfrac{\partial C}{\partial P}} < 0.$$

Note also that, as opposed to Keynesian measures, a wage reduction policy reduces the price level, as demonstrated in the following comparative-static analysis:

$$\frac{\partial P^*}{\partial W} = \frac{-(\partial Y_s/\partial W)(1 - \partial C/\partial Y)}{\dfrac{\partial Y_s}{\partial P}\left(1 - \dfrac{\partial C}{\partial Y}\right) - \dfrac{\partial C}{\partial Y}} > 0.$$

For this regime to obtain, the parameters must be such that the equilibrium price P^* is higher than its minimum value $\bar{P}$ and there is an excess supply of labour.

To prove this static-comparative analysis, a simple or multiple regression using either a single structural equation or more complicated systems may be attempted.

A simple elasticity estimate of the effects of aggregate government spending on GNP growth based on the standard St Louis model brought about a series of coefficients for time-series data from the first quarter of 1971 to the last quarter of 1988. The results are given in Table 4.5.

The long-term effect of government spending is 0.322 compared with 0.3092 of money supply ($\dot{M}_1$), while the elasticity coefficient for government spending is 0.3319 compared with 0.4931 of total money supply ($\dot{M}_2$).[11] This explains that government spending is more effective than monetary policy when M_1 is considered, but less effective when M_2 is considered.

TABLE 4.5 $\dot{Y}_t = a + \sum\limits_{i=0}^{4} \beta_i \dot{M}_{t-i}^k + \sum\limits_{i=0}^{3} \gamma_i \dot{G}_{t-i} + \varepsilon_t$

	$\dot{M}1$	1971(I)–1988(IV) $\dot{M}2$	$\dot{M}3$
α_0	0.0532	0.0324	0.0151
β_0	0.1646	−0.5099	−0.2487
β_1	−0.1316	0.6599	0.1449
β_2	0.3291	0.1177	0.1638
β_3	−0.0867	0.7050	0.3306
β_4	0.1957	−0.4796	
γ_0	0.1348	0.1609	0.1249
γ_1	0.0827	0.1038	0.1130
γ_2	0.0277	0.0224	0.0700
γ_3	0.0768	0.0448	0.0966
R^2	0.7242	0.7192	0.6060
RMSE	0.0551	0.0556	0.0645
DW	0.9170	0.9640	0.6730

Notes
A dot over a variable indicates the growth rate of that variable.
$\dot{Y}_t$ = $(Y_t - Y_{t-4})/Y_{t-4}$ after seasonal adjustment.
$\dot{M}1$ = growth rate of money supply (M1)
$\dot{M}2$ = growth rate of total money supply (M2)
$\dot{M}3$ = growth rate of total liquidity (M3)
$\dot{G}$ = growth rate of aggregate government spending, that consists of consumption as well as investment expenditure of central government
RMSE = root mean square error (= $\sqrt{(\Sigma e_i/ni)}$)
DW = Durbin–Watson statistics

The drawback with the equation lies in an identification problem, in that the regression relationship of income growth on both government expenditure and money supply does not explicitly explain whether it is a demand equation or a supply equation. Furthermore, the equation does not take account of negative effects of taxes on income. Therefore, an index is derived explicitly to consider the net demand effects of government fiscal policy:[12]

$$ADGX = GX - TX\left(1 - \frac{NS}{NI}\right)$$

where

$ADGX$ = net demand effect of the government sector
GX = government consumption expenditure and investment
            expenditure
TX = direct taxes plus indirect taxes
NS = national saving
NI = national income.

The approximate effect of the government sector on aggregate demand can be found from the annual percentage change rate of the index. As shown in Fig. 4.1, the net demand effect of government expenditure fluctuated with each peak recorded in 1972, 1975, and 1980, and has remained on a steady slow growth path since 1983. A regression is computed to check the casual relation between government expenditure (GX) and net demand effect $(ADGX)$. The correlation coefficient of $ADGX$ on GX is estimated to be 0.65, showing that a unit change in GX results in a marginal increase of 0.65 unit of $ADGX$. Additional results were obtained for the relationships between the growth rate of government spending (GXR) and the consumer price inflation rate $(CPIR)$, between the growth rate of government spending (GXR) and interest rates (IRR), and between government spending (GX) and GNP as well as exports $(EXPORT)$:

(1) $ADGX = -226.422 + 0.65007GX$ $RMSE = 170.1864$
         (3.812)   (89.160) R^2 $= 0.9974$

(2) $CPIR = 1.7496 + 0.4418GXR$ $RMSE = 5.41615$
         (0.756)   (5.338) R^2 $= 0.6263$

(3) $IRR = 10.7583 + 0.13895GXR$ $RMSE = 4.0591$
        (6.202)   (2.240) R^2 $= 0.2279$

(4) $GNP = -3582.37 + 6.1686GX$ $RMSE = 7383.677$
        (-1.324)   (21.238) R^2 $= 0.9637$

(5) $EXPORT = -846.5196 + 2.64478GX$ $RMSE = 4881.333$
           (-0.473)   (13.773) R^2 $= 0.9178$

Lastly, the welfare (or wealth) effects of government deficit financing were explored by regressing total private consumption expenditure (PCT) on sets of explanatory variables including permanent income (Y), private assets value (W), social insurance assets (SW), government expenditure (G), taxes (T), and government bonds outstanding (D).[13]

TABLE 4.6 Statistics used for the derivation of $ADGX$[a] (billions of *won*)

	CXC	CXK	TX	NI	NS	ADGX
1970	263.4	141.2	280.1	2 355.9	510.7	185.2
1971	344.0	153.8	351.1	2 887.7	556.9	214.4
1972	425.3	151.0	362.0	3 574.3	734.2	288.7
1973	451.7	162.4	428.1	4 549.5	1 233.5	302.1
1974	734.3	205.1	696.8	6 415.7	1 565.6	412.6
1975	1 121.1	338.2	949.6	8 408.7	1 865.5	720.4
1976	1 520.6	429.7	1 317.7	11 418.2	3 451.3	1 030.9
1977	1 919.1	623.0	1 634.8	14 568.9	5 017.1	1 470.3
1978	2 501.2	921.3	2 224.8	19 637.9	7 331.0	2 028.2
1979	3 059.4	1 323.0	2 998.4	24 898.9	8 928.5	2 459.2
1980	4 386.6	1 706.6	3 582.8	29 176.8	8 702.5	3 580.0
1981	5 515.0	1 993.9	4 463.5	35 994.4	10 627.4	4 363.3
1982	6 254.7	2 420.2	5 140.0	40 758.7	13 062.1	5 182.1
1983	6 851.9	2 821.6	6 015.0	47 790.2	17 343.3	5 841.4
1984	7 262.6	3 283.3	6 490.9	54 469.3	20 996.1	6 557.0
1985	8 135.9	3 678.5	7 228.4	60 755.1	23 037.7	7 326.9
1986	9 400.7	3 615.4	8 174.3	70 645.2	30 321.6	8 350.3
1987	10 708.5	4 071.2	9 660.3	82 113.0	39 006.2	9 708.3
1988	12 763.1	4 968.8	12 205.8	96 408.0	47 132.5	11 493.3

[a] CXC = government consumption expenditure; CXK = government fixed capital formation; TX = direct and indirect taxes; NI = national income; NS = national saving; $ADGX = GX - TX(1 - NI)$ where $GX = CXC + CXK$.

Sources: Bank of Korea, *National Income Statistics* (each issue), and *Economic Statistics Yearbook* (each issue).

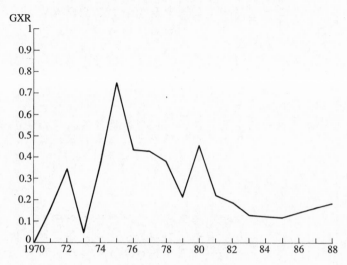

FIG. 4.1 Percentage changes of *ADGX*, 1970–1988

The hypothesis for this analysis is based on the government budget constraint equation; that is, $G - T = dB + dM$, where G is government expenditure and T government tax revenue. If $G - T > 0$, then there is a government budget deficit, which must be financed either through additional bond issue (dB) or through a change in money supply (dM).

Assuming no inflationary revenue financing, that is $dM = 0$, bond issue financing (dB) may or may not have an effect on the real sector. Keynesians emphasize the short-run direct effect of bond financing on real income, but monetarists argue that the wealth effect of government bonds will increase the demand for money by private wealthholders, which increases interest rates, thus resulting in crowding out the effects of government expenditure by a fall in private investment. How much does the amount of government bonds outstanding affect private-sector wealth? A Cochrane–Orcutt estimation method applied to a somewhat modified version of the Blinder–Solow model[14] produced the result that government bonds had about a 48% wealth effect on the aggregate consumption function estimate, while it influenced 58% of wealth effect on per capita consumption function estimation.[15] The outcome implicitly explains that government bondholding is regarded as an asset instead of a future tax burden more vividly by each individual than by consumers as a whole. Keeping such an a priori net asset effect of bonds in mind, the regression analysis of total private consumption expenditure on explanatory variables produced the following equation for the period 1971–88:

$$\ell n\ C = 1.648 + 0.730\ \ell n\ Y + 0.122\ \ell n\ W - 0.596\ \ell n\ SW$$
$$(1.72)\quad (6.60)\qquad\quad (1.56)\qquad\qquad (1.41)$$
$$+ 0.127\ \ell n\ G - 0.125\ \ell n\ T + 0.346\ \ell n\ D$$
$$(1.46)\qquad\quad (1.41)\qquad\quad (1.31)$$
$$R^2 = 0.990;\ DW = 1.60$$

The facts that an empirical testing of marketability of government bonds showed about 48% in aggregate consumption function and about 58% in individual consumption function, and that the marginal consumption expenditure (C_t) with respect to change in government bond outstanding (D_t) shows a positive elasticity of 0.346, well explain the significant existence of the net positive relationship between the amount of government bonds outstanding and its net wealth effect in South Korea.

4.3.4 Banking and Non-Banking Financial Systems

The separation of the Korean economy from the Japanese colonial economic system in the wake of liberation and the subsequent division of the country produced a serious disallocation in the financial system. During the three years preceding the inauguration of the Republic of Korea in August 1948, political instability and crippling economic conditions created a runaway inflation fed by continuous monetary overhang. The existing financial system was not able to cope with the resulting economic situation. This inevitably brought about the drafting of new central and general banking structures, with technical advice from Arthur I. Bloomfield and John P. Jenson of the Federal Reserve Bank of New York. The drafts were passed by the National Assembly in April 1950 and a new central bank, the Bank of Korea, came into being on 12 June 1950, just 13 days before the Korean War broke out. Upon its establishment, the Bank of Korea drafted the General Banking Act, under which commercial banks were to be reorganized. The implementation of the General Banking Act, however, was delayed until August 1954.

After the cease-fire, the primary task facing the central and commercial banks in South Korea was to finance the country's economic rehabilitation. To such an end, the Korean Development Bank was set up in 1954 with capital financed wholly by the government; and in 1956 the Federation of Financial Associations was reorganized into the Korean Agricultural Bank.

Following the military *coup d'état* in 1961, the government re-organized financial institutions to effectively finance the first Five-Year Economic Development Plan launched in 1962. In 1961 a major portion of the equity capital of commercial banks was trans-ferred to the government under the then martial law designed to break the power of *chaebol*[16] and make them subservient to gov-ernment direction. In this it succeeded. The Bank of Korea Act was amended for the first time in 1962 to strengthen government influence over monetary policy.

In the early 1960s, the government also introduced various specialized banks to facilitate financial support for economic devel-opment. They include the National Federation of Agricultural Co-operatives, the National Federation of Fisheries Co-operatives, the Small and Medium Industry Bank, and the Citizens' National Bank. Later in the 1960s, the Korean Exchange Bank and the Korean Housing Bank were established. In 1983, the National Federation of Livestock Co-operatives was added to complete the present set-up of specialized banking institutions.

Meanwhile, the commercial banking system was also restruc-tured to meet the changing needs of the economy. In the early 1980s, the government-held equity capital portion of commercial banks was sold to individuals to liberalize and privatize the banking system. Provincial (local) banks were introduced in the mid-1970s with the prime objectives of supporting regionally balanced devel-opment. As a measure for financial liberalization and internation-alization, foreign banks were allowed to open branch offices, subsidiaries, and representatives in South Korea well before the 1980s. At the end of 1990, there was a total of 66 branch offices of foreign banks doing business in South Korea. As for non-banking financial intermediaries, there has been steady develop-ment both of various non-banking financial institutions and of the securities market. Towards the end of the 1960s, the Korean Devel-opment Finance Corporation, a private development financial institution, later renamed the Korean Long-Term Credit Bank, was incorporated as a long-term financial institution. The Export–Import Bank of Korea was also established to promote financial support for exports and overseas investment.

With the promulgation of the Presidential Emergency Decree in 1972, designed to induce unorganized curb-market funds into the organized financial market, investment and finance companies were introduced to conduct short-term dealings in commercial paper. Unit trusts (mutual funds) and finance companies were also organized from what were formerly pseudo-financial companies so that they specialize in both instalment savings and small loans.

In 1972 merchant banking corporations were introduced, in order to attract foreign capital and to supply medium- and long-term funds for business.

The securities market has also grown rapidly since 1972 as the result of a series of supportive measures to promote investment in securities and to encourage enterprises to go public. Late in the 1970s, various institutional arrangements were made to ensure the sound operation of the market. These include the strengthening of the underwriting function of investment trust companies and the establishment of the Securities and Exchange Commission and the Securities Supervisory Board.

These non-banking financial intermediaries have grown rapidly, thanks to the relatively higher interest rates they are permitted to apply and also to a relatively lower level of control or intervention from the government. Until the end of the 1970s, South Korea's financial markets were heavily regulated by various government controls and interventions. The consequence was a distortion in the allocation of resources and the distribution of income. The South Korean banking system was seriously retarded as heavy government controls had led to a lack of competition, inefficient practices, low profits, and an accumulation of non-performing loans.

Since the early 1980s, however, South Korea has gradually implemented a deregulation and liberalization programme for its financial markets. Yet real sector growth in recent years has not been matched by a comparable development and liberalization of its financial system. In November 1990, bilateral or multilateral issues between South Korea and other industrial market economies still had to submit to government intervention and to constraints on the ability of foreign financial institutions to compete on an equal footing with South Korean institutions. Since a country's financial system is closely related to its macroeconomic policies, the speed of the liberalization would be more or less affected by the current and future situation of the whole economy.

4.4   THE BUDGETS OF NORTH AND SOUTH KOREA

> Suppose one of you wants to build a tower. Will he not first sit down and estimate the cost to see if he has enough money to complete it? For if he lays a foundation and is not able to finish it, everyone who sees it will ridicule him, saying, 'This fellow began to build and was not able to finish.'
>
> *Luke 14: 28–9*

As in other command economies, North Korea's government budget outlays reflect virtually every sector of economic activity, which in turn implies that the state budget shares a very large proportion of GNP, about 50–60% in year-to-year averages. By contrast, the ratio of South Korea's general government (central plus local government) budget spending to its GNP is around 25.0% annually (see Table 4.7). The comparison can be easily made by looking at Table 4.2 for North Korea and Table 4.3(*a*) for South Korea for the periods of more recent years.

In South Korea, the budget's proportion to GNP is expected gradually to expand as financial demands for greater social welfare services work to drive the budget up more rapidly than GNP grows in the years to come. On the other hand, if economic reforms begin to be implemented in North Korea, as is occurring now in most of the former communist bloc countries,[17] growing autonomy in the management of the economy may reduce the size of funds managed through the government budget.

A glance at North Korea's budget reveals that Pyongyang has usually managed to keep the budget in surplus, indicating that expenditure is always less than planned. North Korea has sought to keep economic changes to a minimum and to control in detail any changes that do occur. This policy minimizes any excess expenditure. Thus, the North tries to show the world that its system is 'superior communism', extolling its success in keeping its budget in the black.[18]

South Korea's budgets, on the other hand, displayed a considerable amount of red ink until 1986, although they began to be balanced in 1987. The South Korean authorities regard a balanced budget as a goal, but they realize that financial policy and economic situations can turn even the best budget into a deficit, and they go further in recognizing budgetary imbalance as one means of flexible financial operation. While North Korea clings persistently to the idea of keeping the budget in balance regardless of changes in the economic situation, South Korea keeps a much more flexible approach to its budgetary management.

With respect to the revenue side, South Korea's major sources of revenue are taxes, which account for more than 85% of all revenues. The remaining sources are monopoly profits, income from trust funds, and other non-tax revenue. Of its tax revenue, about 73% came from internal (direct and indirect) taxes and the remaining 12% from customs duties in the late 1980s. Of the internal tax revenue accruing to the central government, the composition of direct taxes *v.* indirect taxes was 61.0% and 39.0% in 1960, 49.5% and 50.3% in 1970, 32.9% and 67.1% in 1980, 36.6% and 63.4% in 1985, and 45.6% and 54.4% in 1989. Table 4.8 shows an international comparison of tax structures. It is interesting to learn that South Korea is more heavily dependent upon indirect taxes than other industrial countries. Indeed, South Korea's tax system centres on the value added tax, which is typical of a state relying on an indirect tax system. By depending more on indirect than direct taxes, the government aims to minimize popular resistance to paying taxes while easily mobilizing the financial resources required to meet national objectives.

In North Korea, meanwhile, all state revenue comes from socialist production activities. Generally, North Korea's tax revenue is called 'socialist financial revenue', which is derived from the transaction revenue (turnover profit), the tax on state industrial profits, the co-operative organizations' income tax, and some other minor sources. The proportion of overall state revenue provided by the transaction revenue and the state industrial profits tax has gradually increased since 1975. Today, the transaction revenue accounts for 65–70% of all state revenue, while state industrial profits contribute about 20%. The transaction revenue fund is defined as 'a pure socialist income category recorded in the budget at a fixed rate of the price as products are manufactured'. The transaction revenue is in effect the sum of the profits of the North Korean economy. The profits generated by the means of production revert

TABLE 4.7 Total government revenue and spending as a percentage of GNP, North and South Korea, 1961–1990

| | South Korea | | | | | | North Korea | | | |
| | Revenue[a] | | | Expenditure[b] | | | Revenue[c] | | | Expenditure[d] |
	Total	Central gov't	Local gov't	Total	Central gov't	Local gov't	Total	Central gov't	Local gov't	
1961	9.5	8.7	0.8	19.2	17.7	1.5	50.4			49.1
1965	8.6	7.2	1.4	11.6	10.1	1.5	54.1			52.6
1970	14.3	13.1	1.2	25.1	21.6	3.5	57.5			55.4
1971	14.5	13.3	1.2	25.3	21.6	3.7	50.6			50.1
1972	12.5	11.4	1.1	26.6	23.1	3.5	50.9			50.7
1973	12.2	10.8	1.4	21.4	17.9	3.5	49.6			47.9
1974	13.5	12.1	1.4	22.6	18.9	3.7	49.2	39.1	10.1	47.6
1975	15.4	13.8	1.6	24.6	21.1	3.5	47.5	37.9	9.6	46.6
1976	16.7	15.1	1.6	24.3	21.0	3.4	47.6	37.2	10.4	45.5
1977	16.7	14.8	1.9	25.0	21.0	4.0	53.0	38.8	14.2	51.7
1978	17.1	15.3	1.9	25.3	19.9	5.4	51.5	43.7	7.8	48.5

1979	17.4	15.5	1.9	26.8	21.0	5.9	49.9	42.5	7.4	51.7
1980	17.9	15.8	2.1	29.0	23.5	5.6	53.5	46.0	7.5	52.9
1981	18.0	15.9	2.0	29.2	23.5	5.8	56.7	47.8	8.9	55.7
1982	18.2	16.1	2.1	28.7	22.1	6.2	55.4	47.0	8.4	54.3
1983	18.5	16.3	2.3	27.3	20.4	6.8	53.1	44.9	8.2	52.3
1984	17.7	15.6	2.2	27.9	20.4	7.1	55.8	47.1	8.7	55.5
1985	17.3	15.2	2.1	26.2	19.2	6.9	56.6	48.8	7.8	56.4
1986	17.0	15.0	2.0	25.9	19.0	6.6	57.7	49.2	8.5	57.4
1987	17.5	15.5	2.1	25.0	18.1	6.8	59.4	50.9	8.5	58.9
1988	17.9	15.4	2.5	23.2	18.0	5.2	60.6	51.5	9.1	60.2
1989	18.5	15.0	3.5	27.8	19.7	8.1	62.0			61.6
1990[e]	17.1	13.9	3.2	27.9	18.6	9.2	64.3			64.3

[a] Revenue consists of taxes (internal taxes and customs duties), monopoly profits, non-tax revenues, trust funds, etc.

[b] Expenditures data for South Korea partly overlap with data in Table 4.3(a).

[c] Revenue of North Korea consists of transaction turnover profits and others (see text).

[d] Expenditure data for North Korea partly overlap with data given in Table 4.2.

[e] Preliminary.

Sources: For South Korea, EPB, Major Statistics of Korean Economy, each issue; for North Korea, Chosen Chungang Nyungam, each issue, and Table 3.11(a) above.

TABLE 4.8 International comparison of tax structures, 1960–1989 (%)

	South Korea[a]		Japan		USA		UK		West Germany	
	Direct taxes	Indirect taxes	Direct taxes	Indirect taxes	Direct taxes	Indirect taxes	Direct taxes	Indirect taxes	Direct taxes	Indirect taxes
1960	29.8	70.2	54.3	45.7	83.2	16.8	53.4	46.6	47.3	52.7
1965	42.9	57.1	59.2	40.8	84.2	15.8	53.9	46.1	48.6	51.4
1970	43.5	56.5	66.1	33.9	87.1	12.9	55.2	44.8	47.9	52.1
1975	39.5	60.5	69.3	30.7	88.4	11.6	63.7	36.3	52.8	47.2
1980	36.9	63.1	71.1	28.9	90.6	9.4	59.2	40.8	52.2	47.8
1981	37.9	62.1	70.1	29.9	87.9	12.1	59.7	40.3	51.2	48.8
1982	37.5	62.5	70.8	29.2	89.5	10.5	58.8	41.2	51.8	48.2
1983	35.7	64.3	71.0	29.0	88.8	11.2	57.5	42.5	50.4	49.6
1984	38.1	61.9	71.5	28.5	88.9	11.1	56.5	43.5	50.5	49.5
1985	39.3	60.7	72.8	27.2	90.0	10.1	56.1	43.9	52.7	47.3
1986	38.9	61.1	73.1	26.9	90.0	9.1	54.2	45.8	53.1	46.9
1987	41.1	58.9	73.3	26.7	91.5	8.5	53.9	46.1	52.8	47.2
1988	44.9	55.1	73.4	26.6	—	—	—	—	—	—
1989	49.8	50.2	72.1	27.9	—	—	—	—	—	—

[a] Including both central and local government tax revenue.

Sources: Ministry of Finance (Seoul), *Tax Statistics* (1989); Japanese Ministry of Finance, *Finance Monthly* (May 1989).

to the appropriate revenue category of the budget. The transaction revenue is, in fact, an integral part of product pricing in the North Korean economy.

The state industrial profits tax comprises the planned or excess profits generated by the socialized industries. In the first instance, the revenue from the state industrial profits tax is returned to the factory or firm that generated it, where it is utilized to increase production, to replenish working capital, and to meet other needs. Any left-over revenue then reverts to the state. These industrial profits in the socialist economy are equivalent to a kind of corporate profits tax in a market-oriented economy.

In North Korea, a tax is levied on the profits of co-operatives producing commodities or engaged in fishing or agriculture. The overall proportion that this tax contributes is very small and is gradually declining. Other minor revenue sources include fines and penalties, depreciation recovery, aid from abroad and the like, which comprise about 7%–8% of total tax revenue.[19]

Regarding budget outlays by function in both states, South Korea spent about 27.1% on national defence, 34.2% on education and social services, and 20.8% on economic development in 1988. From a historical perspective, defence expenditure has remained rather constant over time, while social welfare and education expenditure has been gradually increasing in recent years, as the South Korean people's demand for social services rises. Economic development expenditure has recorded a gradual decline.

North Korea, on the other hand, spends about 67.0% on economic development, 19.0% on social services, and 12.2% on defence. As already explained, the elasticities of North Korea's expenditure, by function, exceed those of South Korea with the exception of administrative management. It is rather surprising that the centralized command economy in the North has a lower expenditure income elasticity in general administration than the decentralized market economy in the South. Perhaps the structure of general administration is much simplified in cost-saving ways in the North as contrasted to the rather expanded structure in the South. Another notable point is that North Korea is spending the largest proportion of its budget on economic development, mainly of heavy industry. It is known that the state provides most of the funding for socialized industry, government agencies, and supplemental outlays for fixed capital investment (basic construction investment) and liquid asset funds. The category 'economic development' also includes a variety of other expenditures related to the economy, such as price support funds, emergency funds, expenditure on expanding stockpiles, reserve funds, and the replenishment of trust funds. Of these, the greatest share goes to fixed capital investment, which grew from 44% in the early 1960s to 60–65% in the 1970s and 1980s.

North Korea's state budgetary expenditure for 1990 was planned to be 6.9% above the 1989 level, and a big part of it was to go to the expansion and strengthening of the heavy industry sector. The construction of power stations, coal and ore mines, and metallurgical bases are to be aggressively pushed ahead, with investments 7.5% above the previous year's level. A marked increase of investment was also budgeted to be poured into the machine-building, electronics, and automation industries in 1990, compared with 1989's emphasis on development of the chemical industry, railway transport, light industry, agriculture, and housing.[20]

In 1990, North Korea allocated about 12.1% of total budget outlays for defence, whereas South Korea poured about 30.2% into the military sector. According to the official statistics, North Korea's military expenditure has been far less than that of South Korea since 1972 (see Table 4.1(*b*)).

If the defence budgets of both countries were translated into percentages of each GNP, they represent about 17.2% in 1970 and 7.4% in 1989 for North Korea and 3.7% in 1970 and 4.4% in 1989 for South Korea. In 1989, for example, the absolute size of South Korea's economy in US dollars was about four to six times as large as that of the North, depending upon the exchange rates being applied to North Korean GNP. Consequently, South Korea's defence spending of 4.4% of its GNP is equivalent to 17.6–26.4% of North Korea's GNP, while the latter is actually spending only 7.4% of its GNP for defence. This gap between fact and reality must pose a significant psychological burden and threat to the North, and one exacerbated by its currently stagnant economy.

In terms of aggregate military expenditure expressed in US dollars, both states spent almost equal amounts until 1981; since then, the South has led the North. Table 4.9 shows the comparisons along with other estimates by the International Institute of Strategic Studies (IISS), the Stockholm International Peace and

Research Institute (SIPRI), and the US Arms Control and Disarmament Agency (ACDA).

Some observers do not rule out the possibility that North Korea may be concealing a considerable proportion of actual military expenditure in other budgetary sectors, particularly that of economic development. In any economy where a large number of heavy industries are more or less connected with the military sector, such an assertion must have some validity. In fact, about 49.6% of total investment was directed to the mining and manufacturing sector during the 1954–6 period in North Korea. This proportion went up to 51.3% in 1957–60 and to 60.0% thereafter. The ratio between heavy and light industry was 81.1% and 18.9% in 1954–6, and 82.6% and 17.4% in 1957–60. Since then, light industry has consistently occupied about 28% of annual investment, while heavy industry takes up 72%. Therefore, North Korea has continuously concentrated on both heavy industry and national defence, as it wants to maintain a balance between 'tractors' (economy) and 'tanks' (defence).

In South Korea, meanwhile, the mining and manufacturing sector has grown from 16.2% in 1962 to 22.4% in 1971 to 31.3% in 1981 and to 33.2% in 1988. Light industry accounted for about 73.2% of this sector until 1962, when it was overtaken by heavy industry which came to occupy about 55.4% by 1985 and 60.0% in 1990. Approximately a third of heavy industry is very likely related to the military sector in the South if the amount of military procurement is taken into consideration. Thus, heavy industry has definitely something to do with national defence in both states. A conservative estimate indicates that North Korea is currently devoting about two-thirds of heavy industry for military uses, as compared to one-third in the South.

Defence spending can be subjected to two separate, but related, lines of enquiry. The first is to analyse military spending using a theoretical framework that stresses factors influencing defence output and expenditure decisions. This line of research usually focuses on estimating military expenditure demand functions.[21] The second is related to the pure public good attribute of national defence, i.e. the optimum size of defence goods which will maximize social welfare.

The trade-off between defence goods and non-defence goods can be conceptually explained by Wagner's diagram[22] as follows.

TABLE 4.9 Aggregate military expenditure, North and South Korea, 1957–1990 (US$ million)

	South Korea[a]	North Korea[b]		IISS[c]		SIPRI[d]		ACDA[e]	
		(A)	(B)	South	North	South	North	South	North
1957		47.1	25.7						
1960	105.9	50.8	27.7						
1961	119.2	48.7	26.6						
1962	161.9	—	—						
1963	106.5	—	—						
1964	96.4	—	—						
1965	106.8	231.7	126.4						
1966	131.2	—	—						
1967	155.8	1000.2	467.0						
1968	200.6	1299.5	606.8						
1969	252.3	1304.2	609.0						
1970	304.0	1550.7	724.1	333	746	325	745	491	576
1971	382.1	1765.6	930.4	411	849	385	892	654	757
1972	481.8	1131.6	532.2	427	443	435	1045	748	1025
1973	473.9	1153.5	542.5	476	620	411	1068	896	1084
1974	691.8	1622.1	659.8	558	770	1465	1790	982	1368
1975	859.6	1941.9	909.4	719	770	1695	2143	1330	1243
1976	1 389.8	2151.4	960.6	1500	878	2435	2366	1967	1305
1977	1 794.5	2183.1	974.8	1800	1000	2892	2409	2347	1253
1978	2 474.5	2520.7	1260.3	2600	1030	3604	2694	2971	1310

			3200	1200	3363	2946	3089	1315	
1979	2 693.3	3130.7	1469.2						
1980	3 258.0	3197.9	1617.8						
1981	3 222.4	3271.0	1700.2						
1982	3 423.7	3353.5	1534.4						
1983	3 658.5	3461.5	1619.6						
1984	3 655.9	3182.5	1618.2						
1985	3 790.6	3677.9	1619.5						
1986	4 304.6	3925.3	1795.4						
1987	5 185.5	4224.7	1855.7						
1988	6 686.0	4109.2	1839.4						
1989	10 449.2	4129.8	1907.6						
1990	9 721.3	4494.1	2074.2						

[a] For calculation of 1990 estimate, the exchange rate is assumed to be W715.3 per US dollar, which existed in June 1990.
[b] (A) is based on official exchange rate, while (B) is based on trade exchange rate.
[c] International Institute of Strategic Studies.
[d] Stockholm International Peace Research Institute.
[e] US Arms Control and Disarmament Agency.

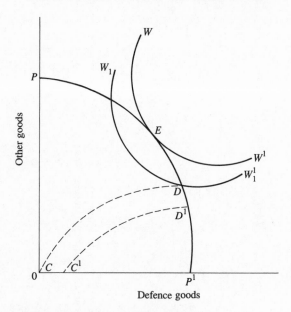

FIG. 4.2 Social optimum with national defence

Fig. 4.2 shows a society's production possibility curve PP^1, and social preference curve WW^1. In the absence of international trade, society's optimum allocation of resources between defence and non-defence sectors will be point E, where PP^1 and WW^1 meet. However, in light of the public good attribute of national defence, the production possibility curve PP^1 is not necessarily equal to society's consumption possibility curve; for, in the absence of national defence, the total amount of non-defence goods produced cannot possibly be enjoyed fully by all people in the community when a foreign invasion takes place. There must exist a certain degree of complementarity between defence goods and non-defence goods spending. The *raison d'être* of defence goods expenditure lies here. With some amount of defence goods being produced with a country's given resources constraint, society's consumption possibility curve will be like CDP^1. In this case, the optimum equilibrium will occur at point D, which is lower than E, indicating a lower social welfare level upon the introduction of defence goods.

As military threats rise from abroad, the *CD* curve shifts to the right to, say, C^1D^1. As a result, the country will end up with more defence expenditure and less social welfare consumption, as indicated by $W_1W_1^1$. It must be pointed out that, the more a state is committed to military spending in a world of resource constraints, the less is social welfare consumption, since less resources will be available for the production of non-defence goods.

This is well demonstrated by the relatively low social welfare expenditure in both North and South Korea. The North's social welfare expenditure category corresponds to both social welfare and educational expenditure in the South. Education, health, physical education, social welfare, social insurance, academic research, and day care expenses are all included under social welfare and cultural policy expenditure in the North. Within this function, North Korea places particular emphasis on education and health. Nevertheless, the North has allocated only about 20% of its budget to this category (see Table 4.1(*a*)). While North Korea is steadily reducing expenditure on this function as a proportion of the total budget since the 1970s, South Korea has recently begun to increase expenditure on this category (social welfare plus education in Table 4.3(*b*)). The South is currently spending about 30% of the budget for this function, which is, however, still very low when compared with such spending by other industrial countries: 46.4% in the United States; 40.8% in Britain; 59.0% in Sweden; 69.8% in West Germany; and 67.1% in France in 1982.[23]

Lastly, it is worth considering briefly the financial structures of central and local government in North and South. In South Korea, local governments were still relying heavily on the national budget being entirely governed by the central government in November 1990 (when this book was being written).

The extent to which local governments were depending on financial support from the central government amounted to 6.7% of the general budget in 1976, 6.4% in 1980, and 8.0% in 1985, and was estimated to be about 9.3% in 1990. This degree of dependence also amounted to 11.2% of total internal taxes in both 1976 and 1980, 13.6% in 1985, and approximately 13.3% in 1990. In addition, about 8.0% of the general account budget annually is being supplied to local governments for educational expenses.

North Korea, meanwhile, has a communist system of central budget and local (or regional) budgets, which taken together constitute the People's (State) Budget. The central budget targets the entire society both in procuring budget income and in allocating budget expenditure, while the local budgets assign local revenue and allocate expenditure for local administrative agencies, local factories, industrial plants, service facilities, etc.

In both North and South Korea, the central budgets overwhelm the local budgets. The comparative compositions are given in Table 4.7 above.

4.5   CONCLUDING REMARKS

This chapter has dealt with the constituent elements and comparative aspects of public finance of North and South Korea.

In a centrally planned system like North Korea, fiscal tools and the budget are used for purposes of resource allocation, but this is effected mainly through the actual transfer of resources rather than via monetary signals and levers, as is generally the case in a market-oriented economy like South Korea. North Korea uses physical quantities as its main planning tool and utilizes output targets as its main indicator of success and generator of incentives, while prices, wages, and interest rates are fixed and seldom altered. However, the production sector is controlled not only by the physical plan, but also through a financial plan formulated by the North Korean Ministry of Finance.

The share of the state budget of GNP amounts to about 60% in the North, suggesting that such finances virtually support a large part of the country's economic activity. By contrast, the South Korean budget's proportion of the GNP is about 20%, though this is likely to expand gradually as the financial demands for increased social welfare services raise the budget faster than GNP growth.

The elasticities of revenue and expenditure with respect to income are very high in both aggregate and category data in North and South Korea. Inasmuch as the state budget serves to allocate national resources, budgeting is crucial to the economies of both North and South.

North Korea has maintained a balanced budget, unlike other communist countries. South Korea's budget, on the other hand, was in deficit until 1986, although its budget began to be balanced in 1987.

While welfare expenditure is low in both countries, each is allocating a considerable proportion of its resources to defence and defence-related activities. Industries linked to the military sector must have enjoyed tremendous monopoly positions in both North and South Korea.

As far as non-budgetary financial systems are concerned, the role of bank and non-banking financial institutions is limited in North Korea. Financial institutions manage the accounts of state enterprises and provide them with credit in accordance with the state plans. Each physical allocation of an input or a supply target to any enterprise is backed by a special bank account. Neither a specific plan's account nor the credit provided can be used for anything other than the intended purpose; nor in most cases are any residual balances left in such accounts unless permitted by the Workers' Party or the President. All inter-enterprise transactions or transactions with the state are made by transferring balances between accounts, and all profits and surpluses not predesignated for enterprise use are transferred to the state. Such operational rigidity was universal in centrally planned economies including the Soviet Union.[24]

In the South, however, financial institutions, though still under some government regulation, engage in basically all types of profit-motivated financial businesses, which in turn contribute to the functioning of the market economy.

NOTES TO CHAPTER 4

1. *Chosen Chungang Nyungam* (North Korea's Central Yearbook), *1949*, 117.
2. Ibid. 117.
3. The tax system of 1948 was subsequently changed or modified when the North socialized all means of production and property.
4. Peter T. Knight, 'Economic Reform in Socialist Countries: The Experiences of China, Hungary, Romania and Yugoslavia', World Bank Staff Working Paper no. 579 (Washington, DC, 1983), 3.
5. Households in North Korea, as in other CPEs, are extremely limited in the range of financial assets available to them. Normally they can hold cash, demand deposits, and a few types of time deposits at the state bank. In that sense, money has an additional role as a store of value.
6. Most of this part is taken from Eui-Gak Hwang, 'Trade Policy Issues between South Korea and the United States, with Some Emphasis on the Koreas' Position', paper presented at the Academic Symposium on the Impact of Recent Economic Developments on US–Korean Relations and the Pacific Basin, University of California at San Diego, 9–10 November 1990.

7. The tendency of government expenditures to rise as incomes rise is known as Wagner's Law. Empirical studies of Wagner's hypothesis seem inconclusive, some vindicating it while others refute it. For further details, see Sohrab Abizadeh and Mahmood Yousefi, 'Growth of Government Expenditure: The Case of Canada', *Public Finance Quarterly*, 15: 1 (1988), 78–100.

8. Budget data used for South Korea are from the consolidated general accounts of central government, which include both current and capital revenue and expenditure. The expenditure data are given in Table 4.3(*b*), but revenue data are different from the tax revenues in this table.

9. See Arthur Okun, *Equality and Efficiency: The Big Trade-Off* (Washington, DC: Brookings Institution, 1975).

10. The controversy between Keynesians and monetarists centres around the relative importance of key instruments of macroeconomic management. The Keynesian view exaggerates fiscal dominance and monetary utility, while monetarists take the opposite view concerning unemployment. The crowding-out effect refers to the case that in a world where the *LM* curve is inelastic and the *IS* curve is elastic, fiscal policy results only in changes in nominal variables, not real variables.

11. See Eui-Gak Hwang, 'The Macroeconomic Effects of Government Spending in South Korea, 1971–1988', *Korean Economic Studies*, 4: 1 (1990).

12. The index was borrowed from Rutaro Komiya and Kazuo Yasui, *Japan's Macroeconomic Performance since the First Oil Crisis: Review and Appraisal*, Carnegie–Rochester Conference Series on Public Policy no. 20 (Amsterdam: North-Holland, 1984).

13. This model is similar to models developed by Feldstein and Kormendi respectively. See M. Feldstein, 'Government Deficits and Aggregate Demand', *Journal of Monetary Economics*, 9 (1982), 1–20; and R. C. Kormendi, 'Government Debt, Government Spending, and Private Sector Behavior', *American Economic Review*, 73 (1983), 994–1010.

14. See A. S. Blinder and R. M. Solow, 'Does Fiscal Policy Matter?' *Journal of Public Economics*, 2 (1973), 319–37.

15. Jung Hae-Koo, 'The Bond Financing Effect of Budget Deficits and Optimum Size of Government Bond Financing' (in Korean), Ph.D. dissertation (Korea University, 1989).

16. *Chaebol* is a Korean expression which can be defined as a financial clique consisting of varied corporate enterprises engaged in diverse businesses and typically owned and controlled by one or two interrelated family groups. Among the more salient characteristics of the typical *chaebol* are (1) family control and management; (2) paternalistic leadership; (3) centralized planning and co-ordination; (4) an entrepreneurial orientation; (5) close business–government relations, and (6) strong school ties apparent in hiring policies (see Richard M. Steers, Yoo Keun-Shin, and Gerardo Ungson, *The Chaebol: Korea's New Industrial Might* (New York: Harper & Row, 1989), 37.

17. Even the former Soviet Union is now moving gingerly towards capitalism. On 16 October 1990 Mikhail S. Gorbachev sent the Soviet Parliament a compromise economic reform programme calling for dismantling of the state's economic monopoly. Major contents of the programme were as follows:

1. *Prices* State control would be phased out. By 1993, only essential consumer goods such as bread, meat, dairy products, medicine and transportation would be controlled. Republics could impose controls, free prices, or rationing.

2. *Property* State-owned factories and farms, and other property such as housing, would be sold off to private buyers, entrepreneurs, collectives and stockholders.

3. *Entrepreneurs* 'Freedom of economic activity' would be established as a right.

4. *Farming* State management of farming and quotas would be abolished. Collective farms could decide whether to dissolve or continue. Republics would have the right to give any member of a collective farm the right to claim and farm land.

5. *Power-sharing* The central government would retain control of transportation, communication, defence industries, energy, credit and monetary policy, customs, and regulation of prices on some raw materials, production, goods, and services. Economic policies would be overseen by a committee with representatives of the republics (see *New York Times*, 17 October 1990).

18. Hyong-Muk Yeon, premier of the DPRK State Administration Council, addressed a national meeting of financial and bank clerks held in Pyongyang on 13 September 1990 as follows: 'During the 7-year plan period, we used the state funds most effectively by strengthening the basic structure of heavy industry and reinforcing its might and by promptly developing light industry and agriculture. We correctly ensured the balance of revenue and expenditures in the state budget, and constantly increased the revenue and expenditures in the state budget according to the rapid growth of socialist construction . . .'

19. Han-Sung Yu, 'The Arms Race *v.* Nations' Welfare' (in Korean), *Wolgan Chosen* (monthly magazine) December 1989.

20. See the summing up of the fulfilment of the state budget of the DPRK for 1989 and the state budget for 1990, which was announced through Pyongyang KCNA by the Ninth Supreme People's Assembly (FBIS-EAS-90-102, 25 May 1990, 26).

21. The demand for military expenditure in each country can be modelled as follows:

$$\lambda n M E_t = \alpha_0 + \beta_0 \lambda_n C_t + \beta_2 \lambda_n \left(\frac{PM}{PC} \right)_t +$$
$$\beta_3 \lambda_n S S_{t-1} + \beta_4 \lambda_n S N_{t-1} + \beta_5 \lambda_n M E_{t-1} + U_t,$$

where $U_t = P_1 U_{t-1} + P_2 U_{t-2} + l_t$; ME = demand for military expenditure; C = civilian output; PM/PC = relative price of military to civilian output; SS = share of military expenditure in output in the South; SN = share of military expenditure in output in the North; and λ_n = natural logarithms operator. For other alternative specification of models, see Rodolfo A. Gonzalez and Stephen L. Mehay, 'Publicness, Scale, and Spillover Effects', *Public Finance Quarterly*, 18: 3 (1990).

22. R. E. Wagner, *Public Finance* (Boston: Little, Brown, 1983), 391.

23. World Bank, *World Tables*, 1988–89 edn. (Baltimore: Johns Hopkins University Press, 1989).

24. See Gur Ofer, 'Budget Deficit, Market Disequilibrium and Soviet Economic Reforms', *Soviet Economy*, 5 (April–June 1989), 107–61.

5

External Relations

Accordingly, whatever you would have people do for you, do the same for them; for this covers the Law and the Prophets.

Matt. 13: 30

It is an irony to note that the post-war drive for economic development and the establishment of external relations of both Koreas has been fed mainly by a strong and virulent nationalism and by the thriving antagonism between the two governments. It would seem that the more dictatorial and totalitarian the auspices under which development takes place, the greater is the accent on nationalism and power at the expense of consumer sovereignty. This tendency is pronounced in both countries, where policy-makers, exercising practically unlimited power, have tried to impose their own preferences regarding present and future consumption and external relations.

The two camps have inadvertently been 'complementary' to each other in the sense that each side has probably been the major determinant of the other side's external relations and internal policies. If some third nation recognized the South, for example, the North would demur to recognize that nation in the hostile environment of diplomatic competition that existed.[1] Economic development in the South is in part a response to the North, or is at least given an urgency that would not exist without the threat from the North.[2] Rigid political and social controls—indeed, the merciless purge of dissidents and repression—have been continuously justified by both states' rulers as necessary because of threats from 'the other side'.

However, external trade regimes have developed differently in both states, basically in accordance with the differing needs and development philosophies perceived by each side. The pattern of foreign trade certainly involves economic and political systems, not to mention the political and ideological preferences of policy-makers in each camp—although the choice of trade partners or

countries has been selectively affected by each side's oblique interaction with the other.

In general, external economic and political relations may be
linked through alternate or mutually complementary channels such
as exports, imports, foreign aid extended, foreign aid received,
foreign direct investment, tourism, etc. In the case of both North
and South Korea, each is of importance in so far as it affects the
other's dependence on foreign countries. In this chapter, I shall
concentrate on the external transactions of North and South Korea
separately.

5.1 NORTH KOREA

5.1.1 Foreign Trade

North Korea has so far pursued an inward-looking development
policy, in spite of limited domestic markets and resource availability. Until recently foreign trade was relatively neglected in North
Korea under the so-called '*chuche* idea', a thoroughly revolutionary
ideology exhibiting a mixture of socialism and nationalism.

It is easy to detect both economic and political issues in the
drive towards autarky in the North. One political goal is to attain
economic independence from any nation that is believed to be
unreliable in the long run. The Sino-Soviet dispute in the early
1960s alarmed the North Korean leadership, which until then had
relied substantially upon its foremost allies—the Soviet Union and
China. To combat the South Korean and US threat, real and
imagined, North Korea needed allies, but it felt increasingly the
necessity of maintaining a neutral stance between its two backers.
Furthermore, in order to pursue a neutralist or non-aligned position
between the Soviet Union and China, the North Korean leadership recognized the necessity of enlarging its scope of economic
autonomy. The key to DPRK leadership behaviour from this point
of view is peculiarly to be found in North Korea's communist
nationalism, from which the '*chuche* idea' was introduced on the
occasion of the fiftieth birthday of Kim Il-Sung in April 1962.

As a matter of fact, the striving for autarky has until recently
been a common phenomenon in most communist economies,
reflecting the post-war political tension between communist and
Western nations. Except for China, no communist country had the

wealth of natural resources or the large domestic market of the Soviet Union. Notwithstanding this economic commonplace, most communist economies began to copy the autarkic Soviet development pattern as a common Marxist–Stalinist bloc identity. The volume of trade even between the communist countries has remained at the minimum level, aimed at just supplementing any gap between domestic supply and demand. Such trade reflects natural resource endowments (e.g. oil) more than anything else. The element of comparative advantage based on technology and labour costs does not play a role except for military technology; hence trade is small.

In recent years, important reforms and changes have been occurring in almost every communist economy including the former Soviet Union, but the emphasis on *chuche* is still upheld in North Korea. In saying this I am not denying that there has been an increase in the volume and profitability of foreign trade for the North. However, where foreign trade is concerned, one of the outstanding characteristics of the past has been the intimacy of relations between the DPRK and other communist countries, particularly the Soviet Union. This relationship was simply a function of the close ideological and political bonds, reinforced by South Korea's containment policy of discouraging Western countries' trade with North Korea. Thus, the Soviet Union, China, and East European countries were the major suppliers of military hardware, technology, and other capital goods to North Korea, although recently the DPRK's trade with Western countries has expanded somewhat both in volume and in kind.

The functions of external trade in a communist country can be generalized as follows. First, foreign trade is incorporated into national plans with the aim of supporting the development of the economy, reflecting both the necessity of filling the gap between domestic supply and demand, and a desire to improve the standard of living.

Second, it is hoped that external trade will help to promote a higher level of productivity through the inflow of new technologies from abroad. Also, the relative dependence on imported goods for capital formation cannot be overlooked.

Third, foreign trade of any centrally planned economy has a major function of providing mutual aid and collaboration with other CPEs, as well as being a way of enhancing the interrelations and interdependence with non-communist economies.

In almost all countries, foreign transactions may absorb a significant proportion of domestic output and similarly can be an important source of supplies of raw materials and commodities and foreign exchange resources needed to meet demands that frequently exceed the domestic population's capacity to save. Within such a general nature and framework of foreign trade, North Korea is characterized as upholding the following basic principles of foreign trade:

1. The principle of implementing foreign trade plans under government monopolistic control and management
2. The principle of equity and reciprocity among trading partners
3. The principle of giving priority to mutual trade with communist bloc countries
4. The principle of giving precedence to the export and import of manufactured goods.

With these preconditional trade principles, reinforced by the *chuche* doctrine, the North Korean external trade regime has developed in its own unique and peculiar way over the last four decades. The concept of foreign trade is defined in North Korea's official *Economic Dictionary* as follows:[3]

The foreign trade of a communist country differs from the one of a capitalist which seeks both exploitation through unequal exchange and profit-maximization. Communist foreign trade aims not only to promote mutual economic development on the principle of reciprocity, but also to positively contribute to the construction of the self-independent national economy.

In other words, communist foreign trade aims not to pursue excess profits, but to smoothly promote the exchange of use-value of commodities among trading countries.

Thus, North Korea's foreign trade sector only serves some auxiliary functions: it is simply a tool, not only to supplement domestic demands and supplies, but also to support the development of a national economy whose ultimate goal is to achieve self-reliance and self-sufficiency.

The volume of total trade (export plus import) has expanded from $22.8 million in 1946 to $4776.8 million in 1990, an annual average growth rate of 4.8% over the past four decades. Table 5.1 shows the aggregate trend of North Korea's trade during the period 1946–90.

TABLE 5.1 North Korea's foreign trade, 1946–1990[a] (US$ million)

	Total trade (X + M)	Annual growth rate (%)	Exports (X) (f.o.b.)	Imports (M) (c.i.f.)	Trade balance (X – M)
1946	22.8	—	n.a.	n.a.	n.a.
1949	182.3	—	76.3	106.0	−29.7
1953	73.0	—	31.0	42.0	−11.0
1954	68.3	−6.4	n.a.	n.a.	n.a.
1955	105.3	54.1	45.0	60.3	−15.3
1956	140.3	33.2	65.8	74.5	−8.7
1957	214.8	53.1	100.0	114.8	−14.8
1958	290.0	35.0	135.0	155.0	−20.0
1959	348.0	20.0	113.0	235.0	−122.0
1960	320.0	−8.0	154.0	166.0	−12.0
1961	326.4	2.0	160.0	166.4	−6.4
1962	352.5	8.0	224.0	128.5	95.5
1963	420.8	19.4	190.7	230.1	−39.4
1964	415.6	−1.2	193.4	222.2	−28.8
1965	441.1	6.1	208.3	232.8	−24.5
1966	463.4	5.0	244.2	219.2	25.0
1967	500.0	7.9	260.2	239.8	30.4
1968	582.8	16.5	276.7	306.1	−40.1
1969	696.1	19.4	306.7	389.4	−82.7
1970	805.6	15.7	333.2	472.4	−139.2
1971	866.0	7.5	301.9	564.1	−254.2
1972	1038.7	19.9	399.0	639.7	−240.7
1973	1340.4	29.1	497.7	842.7	−345.0
1974	1980.3	47.7	677.2	1303.1	−625.9
1975	2077.9	4.9	767.3	1310.6	−543.3
1976	1486.9	−28.4	658.1	828.8	−170.7
1977	1500.0	0.9	680.0	820.0	−140.0
1978	1800.0	20.0	866.0	926.0	−60.0
1979	2927.3	62.6	1457.9	1494.4	−36.5
1980	3430.7	17.2	1594.5	1836.2	−241.7
1981	2626.0	−23.5	1071.9	1554.1	−482.2
1982	2835.2	8.0	1258.5	1576.7	−318.2
1983	2554.5	−9.9	1116.9	1437.6	−320.7
1984	2992.7	17.2	1311.7	1681.0	−369.3
1985	3312.5	10.7	1277.4	2035.2	−757.8
1986	3370.6	1.8	1313.3	2057.3	−744.0

TABLE 5.1 (*cont.*)

	Total trade (X + M)	Annual growth rate (%)	Exports (X) (f.o.b.)	Imports (M) (c.i.f.)	Trade balance (X − M)
1987	3980.4	18.1	1471.6	2508.8	−1037.2
1988	4535.8	14.0	1668.6	2867.2	−1199.5
1989	4590.2	1.2	1685.5	2904.7	−1219.2
1990[b]	4776.8	1.7	1857.1	2919.7	−1062.6

[a] Since North Korea does not release trade statistics, data were derived by IMF DOT method; that is, North Korea's imports equal the exports of other countries (in f.o.b.) times 1.1, and its exports equal the imports of its trading partners (in c.i.f.) divided by a factor of 1.1.
[b] Preliminary.

Sources: IMF and IBRD, *Direction of Trade Statistics Yearbook* (each issue); UN, *International Trade Statistics Yearbook* (each issue); JETRO, *Prospects of North Korea's Economy and Trade* (each issue); OECD, *Trade by Commodities* (each issue); National Unification Board (Seoul), *North Korea's Economic Statistics* (1986); Asia Economic Research Institute (Japan), *Asian Affairs* (1991).

The proportion of trade to GNP is generally very low in the North. In 1990 North Korea's GNP was estimated at $57 158.6 million (using the official exchange rate), while its trade volume was $4776.8 million, representing a proportion of about 8.4%. On the whole, the ratio has kept to a single digit over time, which explains why North Korea has consistently pursued an inward-looking industrial development strategy.

In the 1950s and 1960s North Korea's trade expanded slowly and steadily. But after 1970 its trade grew rather more rapidly, though fluctuating irregularly from year to year. The fluctuation has been stronger with imports than with exports.

The DPRK's exports and imports were valued at $333.2 million and $472.4 million respectively in 1970; this grew to $1857.1 million and $2919.7 million in 1990. Thus, its trade deficit rose from $139.2 million in 1970 to $1062.6 million by 1990, an eightfold increase. On the whole, North Korea has intrepidly run a trade haemorrhage with the rest of the world throughout its trade history except for a few years including 1962, 1966, and 1967. Particularly after 1985, the trade deficit began to increase ever more quickly, as the annual rise of imports exceeded exports by more than double.

TABLE 5.2 North Korea's foreign debt, 1987–1989 (US$ million)

Country	1987	1988	1989
1. Western countries	28.0 (54%)	27.3	27.4
Japan	4.5		
West Germany	2.0		
France	1.8		
Sweden	1.38		
Austria	1.0		
UK	0.24		
Switzerland	0.2		
Netherlands	0.19		
Finland	0.16		
Belgium	0.05		
Denmark	0.04		
Others	7.44		
Consortium banks	9.0		
2. Communist countries	24.1 (46%)	24.7	40.4
China	5.8		1.0
Soviet Union	18.3		39.4
3. Total	52.1 (100%)	52.0	67.8

Source: *Survey of North Korean Economy* (Seoul: National Unification Board, 1989), 90; (1990), 25.

Increasing deficits in the current account began to bring foreign debt problems for North Korea in 1974, when outstanding external debt hit the level of about $2.0 billion, which was equivalent to the value of total trade (imports plus exports) in that year. Thereafter, North Korea's foreign debt outstanding continued to snowball, reaching $5.2 billion in 1988 and $6.78 billion in 1989 (see Table 5.2).

In August 1987, following the failure of negotiations to reschedule the foreign debt, North Korea's Western commercial bank creditors declared the country to be in formal default on $750 million of debt which it had incurred in the early 1970s, and began legal proceedings to seize North Korea's assets, including its gold deposits in London. In September the North Korean government signed a draft agreement outlining terms for a rescheduling of the debt over a period of 12 years. In April 1987 the North had begun a joint venture with a mining corporation owned by a Korean resident in Japan to redevelop the Unsan gold mine, in order to repay trade debts to Japan. In June 1988 the DPRK paid its Western commercial bank creditors $5 million under an agreement whereby 30% of the DPRK's outstanding debt of $900 million due to the banks would be repaid by 1991, and the remainder would be cancelled.[4]

Of North Korea's total trade turnover in 1990, about 56.8% was with the Soviet Union, 10.7% with the People's Republic of China, and 10.6% with Japan. More specifically, North Korea exported goods to a value of $1047.4 million to the Soviet Union, $141.5 million to China, and $271.2 million to Japan in 1990. Its imports amounted to $1667.9 million from the Soviet Union, $403.4 million from China, and $193.7 million from Japan as shown in Table 5.3. Trade statistics from the North are very limited and inconsecutive; and they are on-again, off-again in most source books. The country-to-country breakdown of the data in Table 5.3 should not be taken seriously, but merely regarded as estimates to provide some approximate magnitude of transactions with other countries.

One notable point is that up to now North Korea's foreign trade has been overwhelmingly with communist countries, although the proportion with the West has been on the increase recently.

In passing, it must be mentioned that a five-year trade agreement was signed with China in 1986. Under its provisions, China was to export coking coal, gypsum, crude petroleum, and tyres, while North Korea was to provide China with anthracite, cement, steel plate, and other goods. Under an agreement signed in 1988, Iran was to export 2 million tons of crude oil annually to the DPRK. It was reported in 1983 that North Korea aimed to increase the value of its total exports 4.2 times by the end of the decade, compared with the figure for 1980. Under the third Seven-Year Plan, it was proposed to increase total trade turnover 3.2 times by 1993, with particular emphasis on the export of manufactured goods such as machinery and ships.[5]

The composition of North Korea's trade shifted during the 1950s and 1960s; for instance, minerals as a percentage of all exports declined from 82% in 1953 to 7.2% in 1959, while metals increased from 9% to 39.6%. During the same period, machinery and equipment generally comprised 20–30% of imports, while fuels almost doubled from 9.8% to 19.3%. The composition of commodities in

TABLE 5.3 Estimates of North Korea's trade, by country, selected years (US$ million)

	Exports (f.o.b.)						Imports (c.i.f.)					
	1955	1965	1975	1985	1988	1990	1955	1965	1975	1985	1988	1990
Total	45.0	208.3	767.3	1277.4	1668.6	1857.1	60.3	232.8	1310.6	2035.2	2867.2	2919.7
Communist countries	44.8	185.2	483.2	825.3	1179.9		60.0	206.9	719.8	1249.2	2112.7	
Soviet Union	40.8	88.3	210.4	485.1	887.3	1047.4	44.1	89.8	285.7	864.1	1747.0	1667.9
China	—	83.3	179.8	222.5	233.7	141.5	—	97.0	312.5	262.9	345.3	403.4
Poland	—	6.2	18.1	19.3	24.7	14.3	—	4.8	12.5	24.3	31.2	32.4
Czechoslovakia	0.0	7.2	9.3	34.6	30.1	n.a.	4.7	6.0	12.8	33.7	18.3	n.a.
East Germany	0.1	3.3	18.0	25.7	—	n.a.	7.6	4.5	30.6	25.7	n.a.	n.a.
Bulgaria	—	—	14.2	—	18.7	17.7	—	—	11.8	—	19.6	23.7
Hungary	—	3.4	7.6	5.4	14.4	6.6	—	1.6	13.7	5.5	43.4	14.0
Roumania	—	3.5	15.0	18.2	27.6	14.3	—	5.1	24.4	20.2	n.a.	5.1
Yugoslavia	—	—	—	—	3.5	6.9	—	—	—	1.0	0.8	1.8
Cuba	—	—	10.8	12.5	—	n.a.	—	—	15.8	11.8	—	—
Industrial countries	—	—	—	229.2	—	—	—	—	—	375.1	—	—
Japan	—	14.7	64.9	160.9	293.2	271.2	—	16.5	180.6	274.4	250.8	193.7
West Germany	—	—	51.3	56.2	37.0	50.7	—	—	75.6	27.0	43.9	68.7
France	—	—	18.9	3.9	9.0	13.1	—	—	22.7	8.6	17.5	12.2
Italy	—	—	2.4	1.0	2.4	4.3	—	—	12.2	14.4	21.9	21.7
Spain	—	—	1.0	2.2	3.6	n.a.	—	—	0.4	2.7	4.4	5.8
Holland	—	—	4.8	0.2	0.3	11.3	—	—	1.3	9.5	12.6	n.a.
Switzerland	—	—	0.5	0.2	1.4	n.a.	—	—	10.0	3.1	5.7	n.a.

UK	2.9	2.6	—	1.3	n.a.	—	1.7	3.6	6.4	n.a.
Sweden	0.6	0.5	—	0.8	n.a.	—	66.8	1.3	2.7	n.a.
Austria	0.9	0.2	—	10.0	0.7	—	24.3	24.5	21.9	24.2
Others[a]	—	1.3	—	12.6	—	—	—	6.0	36.6	—
Developing countries	—	222.9	—	—	—	—	410.9	—	—	—
Hong Kong	6.9	18.3	—	29.9	25.8	—	1.7	55.6	131.0	118.4
Thailand	0	10.3	—	28.8	23.0	—	—	9.8	5.8	13.8
Malaysia	2.1	0.4	—	16.6	18.4	—	—	1.4	5.9	0.5
Singapore	12.0	6.1	—	59.1	n.a.	—	8.0	24.9	74.5	n.a.
Indonesia	—	2.8	—	8.7	47.4	—	—	14.7	14.4	37.4
Bangladesh	0.2	19.7	—	17.4	15.1	—	—	5.1	2.7	0.5
India	4.8	25.7	—	42.8	47.4	—	4.8	4.4	28.4	37.4
Iran	0.8	2.4	—	0.1	n.a.	—	—	—	0.1	—
Kuwait	—	77.7	—	—	n.a.	—	—	281.3	—	—
Saudi Arabia	8.1	3.3	—	0.1	n.a.	—	—	—	2.1	—
Yemen	—	7.0	—	10.6	n.a.	—	—	—	—	—
Egypt	24.7	13.3	—	7.9	n.a.	—	8.9	0.7	12.8	n.a.
Zimbabwe	—	0.1	—	0.1	n.a.	—	—	0.3	3.9	—

[a] Includes communist countries, industrial countries, and developing countries that are not clearly specified in the trade statistics.

Sources: IMF and IBRD, *Direction of Trade Statistics Yearbook* (each issue); UN, *International Trade Statistics Yearbook* (1986); JETRO, *Prospects of North Korea's Economy and Trade* (February 1991); Asia Economic Research Institute (Japan), *Asian Affairs* (1991); National Unification Board (Seoul), *North Korea's Economic Statistics* (1986).

North Korea's trade has varied over time, depending upon its trading partners. On the whole, the chief exports are currently metal ores and products, some manufactured goods such as machinery, and marine products, while the chief imports include machinery, tools, petroleum products, fertilizers, tyres and tubes, electronic products including refrigerators, aluminium products, communication tools, combustion engines, etc.

In sum, North Korea's trade policy has evolved in close connection with its economic policy towards industrialization. Imports are largely to meet the demands made in accordance with priorities given to industrial development plans, and exports are promoted to secure the hard currency that is needed to pay for imports. Trade policy regimes are very limited and rigid, since North Korea's decision-making process does not rely on either the price mechanism or quantity signals in the allocation of resources. In spite of its avowed principles of self-reliance and self-sufficiency, however, North Korea's foreign trade has steadily expanded along with the increase in its modernization and industrialization efforts. Nevertheless, its trade dependency ratio (the proportion of the value of imports and exports to GNP) has remained pretty constant over time, at about 9.0% in the 1950s and 1960s and 8.9% in the 1970s and 1980s.[6] This steady ratio indicates that the foreign trade sector has been strongly regulated so as to keep up with domestic industrialization and growth paths.

In general, the evolution of North Korea's trade policies can be classified into three stages. The first stage (1957–60) was related to its export drive to offset decreasing economic aid from both the Soviet Union and other communist countries. North Korea singlemindedly exploited its labour and other available resources to expand export markets and commodities during this period. It also began not only to shift from exports of raw materials to semi-manufactured goods but also to enlarge its group of trading partners to include Japan and Middle Eastern and African countries. The second stage (1960–74) is characterized as having resulted in more imports and foreign debts that were increasingly incurred for the expansion of heavy industry and defence since the beginning of the first Seven-Year Plan period. The trade policy during this period could not help but be directed towards a diversification of exports as well as a growth in export income through increased production of value-added goods.

North Korea's trade relations with non-communist countries began to expand relatively quickly from the early 1970s. In fact, it became necessary to expand ties with Western countries in order to upgrade and modernize its production equipment and machinery. Modern industrial plant and technology have been imported mostly from Japan and Western countries since the early 1970s.

The third stage (1975–90) began with a rather bitter experience of having to reschedule foreign debts in the late 1970s. Meanwhile North Korea became preoccupied with the conviction on the one hand and the difficulty on the other of having to open up its self-reliance-oriented economy in the midst of the repugnant muddle inexorably taking place in most communist economies.

At the time of writing North Korea seems to be seriously concerned about the possible repercussions of the implosion of other communist economies.

5.1.2 *External Economic Co-operation*

From the start, the Democratic People's Republic of Korea received substantial economic and military aid from the Soviet Union and China. Up to 1984, North Korea was known to have received $1278.4 million of aid, and $3470.1 million of loans from abroad, which it invested in economic development and defence. Of total foreign capital imported, about 73.8% was from the communist bloc countries. The Soviet Union provided about 45.6% of total foreign aid and loans to Pyongyang until 1984 (see Table 5.4).

According to a Radio Moscow report on 18 December 1987, the Soviet Union was giving economic aid to Pyongyang even then.[7] This included:

- Expansion of the Kim Chaek steelworks, to increase capacity to 2.5 million tons per year
- Three new coal mines at Anju, to double annual capacity to 4 million tons
- North Korea's very own Chernobyl, a nuclear power plant with a 1700 megawatt capacity
- The East Pyongyang thermal power station
- Two textile mills
- Construction of several satellite tracking stations

Moscow Radio also mentioned a 1987 agreement to establish a joint system for producing machine tools. Both sides were said to

TABLE 5.4 Foreign aid and loan credits received by North Korea, 1949–1984 (US$ million)

	Total	Soviet Union	China	Other communist countries	OECD countries
Up to 1949	53.00	53.00	—	—	—
1950–60	1653.36	713.25	508.50	431.61	—
1961–69	336.68	196.68	105.00	35.00	—
1970	90.00	87.00	—	—	3
1971	267.00	250.00	—	—	17
1972	354.00	150.00	—	—	204
1973	484.00	109.00	—	—	375
1974	520.00	120.00	—	—	400
1975	429.00	186.00	—	—	243
1976	5.60	4.00	1.60	—	—
1978–84	555.85	296.25	258.70	—	—
Total	4748.49	2165.18	873.80	466.61	1242

Source: National Unification Board (Seoul), *North Korea's Economic Statistics* (1986).

be seeking ways to establish more joint venture companies and manufacturing and research systems in North Korea.

According to an EIU Country Report (no. 4, 1988), the UN Development Programme (UNDP) gave North Korea a total of $20.5 million of assistance, most of which was off-the-record, from 1979 to 1987. A new project was also provided by UNDP to assist fruit production and storage at the Pyongyang Fruit Farm in 1988. An agreement to improve rice cultivation was signed in the same year. The total value of the above two contracts amounts to $1.32 million. Other new UNDP subsidy projects initiated in 1988 include the modernization of vegetable farms ($600 000); an electric machinery testing laboratory ($600 000); a soil and plant experiment station ($700 000); and a fish farm ($850 000). Projects like these consolidate the UNDP's role in relatively small-scale, mostly agricultural, projects. That was not the way Pyongyang actually wanted it, preferring ambitious industrial undertakings. But after a UNDP-financed semiconductor factory had run over budget and time

($6.08 million eventually) owing to local incompetence, the UNDP got its way.

It is noteworthy, however, that Pyongyang has survived to be a recipient of technical assistance from the UNDP for an extended period. To celebrate the occasion of the fortieth anniversary of the start of UNDP assistance, a workshop on the development of man was held at the University of National Economy in Pyongyang on 25 October 1990. Attending were the Deputy-Minister of Foreign Affairs, the Vice-President of the University of National Economy, and other involved officials from North Korea, and the resident representative of the UNDP (Mr Henning Karcher) and members of special and affiliated organizations of the United Nations. Topics discussed at the workshop included issues on the role of man in the development of the economy, planning for the development of man, education, health service, and international co-operation.[8]

It is also known that a ten-year agreement (1987–97) on scientific and technological co-operation between North Korea and China was signed in Pyongyang in December 1987. No details were revealed, except for the head of the Chinese delegation stating that China had contributed 1750 different kinds of scientific know-how to further North Korea's development over the previous 30 years. China is also known to have financially assisted North Korea in building the fourth joint Chinese–Korean power station at the Yalu River which was completed in November 1988. The power station, located at the Taepyongman Ri in Sakju county (Liaoning on the Chinese side), has an annual capacity of 770 million kilowatt-hours, with four generators supplying power, two to each country. Earlier similar stations are at Supung, Wiwon, and Unbong, all along the Yalu River.

With the exception of some joint venture projects, North Korea has systematically under-emphasized the amount of foreign aid it has received. This seems a rather idiosyncratic gentility with the country's trumpeted goal of quixotic *chuche* ideology. The other side of the coin is that North Korea has never joined in the club of COMECON (Council for Mutual Economic Assistance), although it has participated in some of its activities.

If North Korea's industry has looked increasingly old-fashioned, despite efforts to modernize technology over the last decade, what is the reason? The ingenuous answer may lie either in the inherent communist system or in the sanguine features of North Korea's self-reliant development policy. In a sense, North Korea may become increasingly a victim of its erstwhile success: the methods that propelled its economy up to now may increasingly become obstacles for the future, unless it decides to observe more openness (*glasnost*) with other countries. Obviously, North Korea is acutely aware of these potential problems. But solving them presents its own risks, and is doubtless the topic of fierce debate behind the scenes.

There have been some signs of change. One was the promulgation in September 1984 of a Joint Venture Law, closely modelled on China's (see Section 5.1.3). There have also been tentative steps towards economic reform, and a degree of decentralization. In late 1985 economic ministries were streamlined into a smaller number of 'commissions' and were further reshuffled a year later. Measures to increase co-ordination between related enterprises— (apparently) on the model of the then East Germany's *Kombinats* —and to increase their autonomy in matters ranging from input sourcing to profit retention and worker bonuses had been carefully attempted. While the effects of these first-stage open policies remain as yet unclear, many external shocks have driven North Korea's reformists to tighten the door again. The Tiananmen Square incident in China in June 1989 and the reform upheavals in Eastern European countries and the Soviet republics must have induced North Korean leaders to give a second thought to their own *glasnost* and *perestroika*, even if already adopted to some extent.

However, the subsequent developments of both Seoul–Moscow's diplomatic normalization and Seoul–Beijing's official trade agreement, signed in October 1990, are likely not to keep North Korea behind its tightly closed door for much longer. Sooner or later, Pyongyang will have to go for inter-Korean exchanges as well as for normalization of relations with Western countries, including Japan and the United States.[9]

North Korea has also been an aid donor to other countries on an unknown but fairly modest scale, mostly African countries, and concerning technical assistance for agriculture. Much larger sums are spent by North Korea abroad on political ends, including the maintenance of diplomatic relations with over 100 countries. Much overseas expenditure is on the propagation of the cult of Kim Il-Sung and more recently for putting forward Kim Chong-Il, elder son of Kim Il-Sung, as the president-designate.

North Korea has also sent educational aid in the form of funds and stipends amounting to a total of ¥39 824 182 433 to the children of Korean residents in Japan (Chongnyon) from April 1957 to February 1990. The 114th instalment of the educational aid fund and stipends was sent by Kim Il-Sung on 14 February 1990, bringing the total to ¥144 million.[10]

Table 5.5 shows the educational aid funds and stipends sent by President Kim Il-Sung to Korean residents in Japan from the 1950s to the mid-1970s, and the total amount up to February 1990 and to April 1993. However, North Korea has never revealed how it raised the funds that have been sent to its compatriots living in Japan.

5.1.3 Joint Venture Companies

Faced with decreasing economic assistance from the Soviet Union and China, and an increasing debt default problem with Western countries, North Korea enacted the Joint Venture Law (Hapyoung Bup) on 8 September 1984, mainly to attract foreign capital and advanced technology and to develop overseas markets. The Law consists of five chapters and 26 articles specifying the outline, scope, organization, activities, and distribution of profits of the joint venture companies. Following the enactment of the law, North Korea promulgated the Implementation Rule of the Joint Venture Law on 2 March 1985, the Joint Venture Tax Law and its detailed 'implementation rule', and the Foreigners Income Tax Law on 7 March 1985.

North Korea began actively to approach Western countries, including Japan, France, West Germany, and Hong Kong, even before the enactment of the law—as early as January 1984—to seek possibilities of joint investment projects with foreign businesses. Basically, DPRK's joint venture programme borrowed in part from China's model in opening up its economy to the outside world. The objective of the Joint Venture Law of 1984 is to facilitate the acquisition of foreign capital and advanced technology in a situation of acute foreign exchange shortage and an adverse balance of trade position. North Korean leaders justified the new law as an extension but not a denial of their philosophy of *chuche* and national independence by saying: 'A self-supporting national

TABLE 5.5 Educational Aid Fund and stipends sent by Kim Il-Sung to Korean residents in Japan, 1957–1993

	Date	Amount (yen)		Date	Amount (yen)
1	1957(4)	121 099 086	34	1970(2)	303 121 100
2	1957(10)	100 510 000	35	1970(4)	300 755 000
3	1958(3)	100 000 000	36	1970(9)	297 780 000
4	1958(9)	100 210 000	37	1970(10)	302 850 500
5	1959(2)	176 382 500	38	1971(1)	302 365 000
6	1959(9)	114 654 090	39	1971(4)	301 945 000
7	1960(2)	202 100 000	40	1971(8)	301 910 000
8	1960(9)	217 392 231	41	1971(10)	289 345 000
9	1961(3)	411 066 000	42	1971(12)	302 827 800
10	1962(3)	558 470 000	43	1972(1)	300 825 000
11	1963(3)	401 440 000	44	1972(4)	328 060 000
12	1963(6)	202 770 000	45	1972(7)	342 270 800
13	1963(9)	186 852 644	46	1972(10)	363 425 000
14	1964(4)	303 930 000	47	1972(12)	351 225 000
15	1964(7)	302 940 000	48	1973(2)	350 925 000
16	1964(11)	201 400 000	49	1973(4)	374 176 500
17	1965(3)	302 038 942	50	1973(7)	334 725 000
18	1965(8)	202 020 000	51	1973(8)	321 421 100
19	1965(11)	303 450 000	52	1973(12)	369 684 700
20	1966(2)	303 570 000	53	1974(3)	362 010 000
21	1966(8)	201 860 000	54	1974(4)	595 170 000
22	1966(11)	303 210 000	55	1974(5)	601 425 000

No.	Date	Amount
23	1967(3)	303 420 000
24	1967(7)	201 420 000
25	1967(10)	301 950 000
26	1967(12)	194 246 300
27	1968(2)	305 025 000
28	1968(6)	347 305 400
29	1968(10)	345 783 600
30	1969(2)	299 754 000
31	1969(4)	350 960 000
32	1969(8)	298 261 240
33	1969(10)	247 950 000
56	1974(10)	697 150 000
57	1975(1)	703 450 000
58	1975(2)	752 580 450
59	1975(4)	1 060 122 500
60	1975(5)	608 265 000
61	1975(12)	613 350 000
62–113	1976–89	19 651 926 900
114	1990(2)	144 000 000
1–114 total	1957(4)–1990(2)	39 824 182 433
1–119 total	1957(4)–1991(12)	41 090 000 000
123	1993(4)	113 500 000
1–123 total	1957(4)–1993(4)	41 585 942 433

Sources: Korea Central News Agency; *Chosen Chungang Nyungam* (North Korea's Central Statistics Yearbook) (1976), 438; KCNA Report, 14 February 1990. See also *Rodong Sinmun* (4 April 1992, 15 April 1993).

economy is by no means a closed economy. On the contrary, it calls for development of economic relations on the principles of independence, equality and mutual benefits.'[11]

Thus, five years after China's reforms, but one year before the beginning of Soviet reform, North Korea embarked on what appeared to be a dramatic departure from its advocacy of self-reliance, the results of which are still very uncertain. The Joint Venture Law invited all countries, including capitalist countries with which the North had not yet established diplomatic relations, to invest in North Korea jointly with North Korean enterprises. Pyongyang seemed to be inviting foreign investors to invest in such areas as electronics, automation equipment, metallurgy, mining, energy, chemicals, foodstuffs, construction, and hotels and tourism, by offering numerous benefits. Imports of goods for joint ventures were exempt from tariffs, and profit taxes were also exempt for the first three years, with rates being adjustable from 25% downward depending on profit levels.

For about a year after its introduction, North Korea did seem to devote a great deal of effort to making the joint venture business grow. In September 1985 Kim Il-Sung invited a Japanese businessman, Den Kawakatsu, who had been active in promoting trade between Japan and China before their diplomatic normalization, and asked him to arrange Japanese investment in North Korea.[12]

North Korea claimed that within a year of the law's introduction it had concluded 10 joint ventures and was negotiating for 30 others.[13] The first joint venture project was the Nakwon (Paradise) Department Store, which was jointly established by the Nakwon Trading Corporation of North Korea and Asahi Shosha, a Choch'ongyon (pro-Pyongyang Korean Residents Association in Japan) company. It opened in February 1985 with a main store in Pyongyang and 31 branches throughout the country, selling mostly foreign goods to foreign tourists and the privileged North Korean élite. This project was followed by a French hotel (Yang-Gak-Do International Hotel) and another two Chongnyon (an acronym for Choch'ongyon) projects in 1985.

In spite of the numerous benefits offered, the Joint Venture Law has not met with the universal acclaim of Western countries, perhaps partly because of the North's remaining heavy debt to Western creditors and partly because of limited markets. The joint venture businesses have since progressed, mostly with the participation of loyal North Korean residents in Japan and with a few investment advances from the Soviet Union, China, and other countries. At the end of 1989, the majority of joint ventures were with Chongnyon entrepreneurs which have reportedly signed contracts or reached agreements with North Korea on 102 joint ventures. Of these, 46 were in active operation as of early 1990. Table 5.6 lists the joint ventures in operation in North Korea including those between North Korea and Chongnyon. The total amount of investment in the joint ventures with Chongnyon businessmen were known to amount to ¥11.3 billion at the end of 1989, an average venture being about ¥160 million on the contract basis. The Chongnyon side offered mostly facilities, technology, and cash funds, while the North Korean side provided buildings, land, raw materials, other supplies, labour, and energy.

Amid reports that in May 1990 North Korea abolished the Joint Venture Ministry, which was established under the State Administration Council in 1988 according to the 1984 Joint Venture Law, two new joint ventures were opened in October 1990. Reuter, in a news report from Pyongyang on 22 October 1990, reported that the ministry was abolished two years after its establishment because North Korea's command economy was not negotiable in the international markets where accounts are settled with hard currencies, and because the level of foreign capital investment was extremely meagre.[14]

However, North Korea's Central News Agency (KCNA) reported two cases of new joint ventures on 7 October and 28 October 1990. The first is the Taedoksan Joint Venture Company operated by the Pukchong County Administration and Economic Guidance Committee of South Hamgyong Province and the Ethime Prefectual Association of Traders and Industrialists of Chongnyon. The company has a capacity to process over 4000 tons of marine products a year. All its production processes including fish-drying have been mechanized and automated.[15] The other one is a Chinese joint venture restaurant called the Chongchun Restaurant, built in Kwangbok Street, Pyongyang. The restaurant opened on 27 October 1990 with a ceremony attended by the vice-chairman of the Liaoning Provincial External Economic Trade Commission of China and the vice-chairman of the State Sports Commission of North Korea. This splendid restaurant has an area of more than 10 000 square metres and 2000 seats, serving mainly famous Korean

TABLE 5.6 North Korea's ongoing joint ventures, March 1990

	Venture	Business or commodity	Location	Participating country
1985	Nakwon Stores	Foreign goods	Pyongyang and 31 local branches	Japan
1985*	Yang-Gak-Do Hotel	Hotel	Yang-Gak-Do	France
1985	Changkwang Coffee Shop	Coffee shop	Changkwang	Japan
1985	Dae-Dong Kang Auto Repair Shops	Repair garage	Pyongyang	Japan
1986	Korea–Japan Friendly Hospital (Kim Man-Yoo Hospital)	Hospital	Pyongyang	Japan
1986	Changgwang Soft Drink	Soft drinks	Pyongyang	Japan
1986	Taedonggang Restaurant	Food, beverages	Pyongyang	Japan
1986	Yi Yong-sam Pyongyang Chestnut Co.	Nuts	Songchon County	Japan
1986	6 Feb. Joint Co.	Straw products	Pyongyang	Japan
1987	Tonghaegwan Co.	Food, beverages	Wonsan	Japan
1987	Pyongyang Golf Co.	Golf course	Nampo	Japan
1987	Changgwang Joint Co.	Restaurants, shops	Pyongyang	Japan
1987	Moranbong Joint Co.	Ready-made suits	Pyongyang	Japan
1987	Nakwon Songhwa Clothing Co.	Women's clothes	Pyongyang	Japan
1987	Wolmyongsan Joint Co.	Feldspar mining	Pyongyang	Japan
1987	Susan Joint Co.	Honey	Pyongyang	Japan
1987	Wonsan Patriotic Knitting Co.	Knitted goods	Wonsan	Japan
1987	Hunduk Joint Co.	Tyres	Hamhung	Japan

Year	Company	Product	Location	Country
1987	Namsan Joint Co.	Electric, electronic goods	Pyongyang	Japan
1987*	Hee Chon Machinery Tool Co.	Machine tools	Hee Chon	USSR
1987*	Seaweed Cultivation Co.	Seaweed, oysters, fish	Chongjin	USSR
1987*	Fishery Co.	Marine products	Wonsan	USSR
1987*	Pyongyang Restaurant	Cold noodles	Pyongyang	China
1987	Nungna Joint Co.	Clothing	Pyongyang	Japan
1987	Nakwon Joint Financing Co.	Finance	Pyongyang	Japan
1987	Noana Beauty Research Group	Ginseng cream	Sinuiju	Japan
1988*	Do-Mun Kang Restaurant	Food	Yonkil	China
1988	Pyongyang Silk Joint Co.	Silk textiles	Pyongyang	Japan
1988	Chilbosan Marine Products	Marine products	Chongjin	Japan
1988	Songbong Joint Co.	Small-type transformers	Pyongyang	Japan
1988	Manjang Joint Co.	Herbs	Pyongyang	Japan
1988	Pyongyang Fish Farming	Eel farming	Pyongyang	Japan
1988	Chongchongang Joint Co.	Silk yarn	Huichon, Chagang	Japan
1988	Chongyu Joint Co.	TV, typewriter parts	Pyongyang	Japan
1988	Kwangwang Taedong	Tourist transport	Pyongyang	Japan
1988	Yugyong Joint Co.	Restaurants	Pyongyang	Japan
1988	Pyongun Joint Co.	Restaurants	Pyongyang	Japan
1988	Koryo Joint Co.	Trading business	Pyongyang	Japan
1988	Oaudo Joint Co.	Refrigerator ship operation	Nampo	Japan
1989	Toksan Construction Machinery	Machine repair	Pyongyang	Japan
1989	Pyongyang Packing Materials	Corrugated cardboard	Pyongyang	Japan
1989	Pyongyang Piano Joint Co.	Pianos	Pyongyang	Japan
1989	Myongjon Joint Co.	Medical instruments	Pyongyang	Japan
1989	Kwangpo Joint Co.	Duck feather processing	Changpyong	Japan

TABLE 5.6 (*cont.*)

	Venture	Business or commodity	Location	Participating country
1989	Sinhung Joint Co.	Bikes, electric, electronics	Pyongyang	Japan
1989	Chosen Joint Bank	Banking	Pyongyang	Japan
1989	Chindallae Joint Co.	Women's clothes	Pyongyang	Japan
1989	Manpung Joint Co.	Light industrial goods	Pyongyang	Japan
1989	Chosen Silver & Copper	Commemorative medals	Pyongyang	Japan
1989	Toraji Joint Co.	Stores	Pyongyang	Japan
1989	Taedonggang Auto Repair	Engine salvaging	Pyongyang	Japan
1989	International Chemical	Mineral ores refining	Pyongyang	Japan
1990	Taedoksan Joint Venture Co.	Marine products	Pukchong	Japan
1990*	Korea-China Joint Co.	Restaurant	Pyongyang	China

* Non-Japanese (non-Chongnyon) joint ventures.

Sources: National Unification Board (Seoul), *North Korean Economy* (1989), 102; KCNA (Pyongyang), various reports; also Yu Yong-Ku, 'Joint Ventures: Chongnyon in Japan Provides Technology, Capital', *Chungang Ilbo* (in Korean) (25 July 1990), 9.

and Chinese dishes.[16] North Korea's joint venture advances to the Soviet Union in the areas of oriental herb medicines and farm and fishery products resulted in seven and nine ventures in 1988 and 1989, respectively.

If recent reports concerning the abolition of North Korea's Joint Venture Ministry are true, this would be interpreted as a case of one step back and two forward. North Korea is in no way likely to return to its closed economic policy in the face of strong winds of reform in all communist countries.

As North Korea's uncertain profitability, ideological rigidity, and limited markets have so far made Western investors reluctant to invest there in spite of various favourable offers such as tax-free remittance and income tax exemption, the country has to desperately seek an alternative source of foreign capital. That is to establish a 'special economic zone', an option that had been deliberately avoided for fear that its citizens might become ideologically and economically tainted. Increasing signs of interest in establishing relations with the outside world were therefore revealed when Hapsando, an island near the Sino-Soviet border, was designated a special economic zone in early 1990.[17] In principle, North Korea has established this island on the Tuman River as a special economic zone in order to develop the area as a China–Russia–North Korea 'trade centre triangle'. But its ultimate goal appears to be to open an 'international commerce centre' in the special economic zone, possibly by early 1995.

In addition, North Korea has recently signalled increasingly its eagerness to establish trade and diplomatic relations with the United States and Japan, just as South Korea established full diplomatic and trade relations with the Soviet Union and China in September and October 1990, respectively. As more efforts for joint venture promotion as well as the establishment of a special economic zone can mark a breakthrough for North Korea's economy, so will its gradual openness with Western countries surely contribute towards bridging the chasm between the communist state and Western culture. Moreover, North Korea's *glasnost* policy, although still disguised well enough to minimize its impact on *chuche* politics, deserves attention, because it may provide opportunities for the two Koreas to contact one another and to work towards unification.

5.2 SOUTH KOREA

5.2.1 Foreign Trade

The South Korean economy has often been singled out as a good example of successful implementation of an outward-looking development strategy. South Korea's gross national product has increased more than fifteenfold over the period 1953–89, but the most growth has occurred since its adoption of an export-oriented development policy in the early 1960s.

Politically divided, densely populated, and lacking both domestic savings and vital natural resources, the South Korean economy seemed hopeless in 1961, suffering from almost all difficulties including rampant corruption partly associated with the distribution of huge amounts of US and other foreign assistance which was poured into the country after the war.[18] Before the first overall economic development plan was launched in 1962 by the military government, South Korea remained in the so-called 'vicious circle of poverty', with little industrial base.

The industrial policy, if any, pursued until then may be loosely characterized as a policy of import substitution of non-durable consumer and intermediate goods behind a protective wall of high tariffs and stringent quotas. The annual average growth rate of GNP during the nine-year period (1953–61) after the Korean War was only 3.7% and per capita GNP was no more than $80 per year. Commodity exports remained negligible throughout the period, usually amounting to less than 3% of GNP.

The persistent political instability of those years, as well as the existing inadequacy of the economic and administrative infrastructure, complicated all early economic problems, which in turn became a good pretext for a military coup led by General Park Chung-Hee on 26 May 1961. While the coup had in many ways a negative political and cultural impact on South Korean society, nevertheless, the strength of its backing had strong positive effects on the capacity of the government to pursue its development plans. The military reinforcement of a managerial structure that pervaded society as a whole has resulted in full implementation of central government decisions in South Korea since 1960.

The new military government, aware of the country's limited domestic markets and resources, adopted an outward-oriented strategy in its first Five-Year Economic Development Plan (1962–6), emphasizing the growth of exports as well as import substitution. Ever since its adoption, the policy of export-led industrialization has been significant in helping the economy to expand rapidly, and export growth will continue to play a pre-eminent role in South Korea's development strategy beyond the 1990s.

The essence of the export-led development strategy in its early stage (1962–71) was to make use of the nation's comparative advantage in labour-intensive manufactured goods. In order to achieve the basic goal of export-oriented industrialization and high growth, a package of policy reforms was initiated between 1964 and 1967. The South Korean *won* was devalued by almost 50%, from Won 130 to Won 255 per US dollar in May 1964, and a unitary floating exchange rate system was adopted in March 1965. This measure was followed by about a 100% interest rate rise on bank deposits and loans in September 1965, and as a result deposits doubled every year for the next three years, and were further enhanced by incentives offered to exporters.

Investment grew very fast, and the investment–GNP ratio in current prices rose from 13% in 1962 to 25% in 1971. Growth in real GNP was more than twofold during the period 1967–71, reflecting an annual growth rate of 8.7%.

As the economy rapidly grew, structural changes took place. The secondary industry (mining and manufacturing sector) increased its share of GNP from 16% in 1962 to 22% in 1971, while the share of the agricultural, forestry, and fisheries sector decreased from 37% in 1962 to 27% in 1971.

Such rapid growth and structural change stemmed from the rise in foreign trade. Commodity exports rose from about $55 million in 1962 to $1.07 billion in 1971 on a customs clearance basis, an annual average growth rate of about 40%. Commodity imports have also expanded markedly, from $421.8 million in 1962 to $2.39 billion in 1971. The trade growth was accompanied by a significant change in its composition. The share of primary products exported declined from 73% of the total in 1962 to only 14% in 1971, while that of manufactured goods rose dramatically from 27% to 85%. In imports, the share of capital goods increased greatly from 17% to 28% by 1971.

As the economy entered the initial stage of the third Five-Year Plan period (1971–6), South Korea began to sense that its export markets for labour-intensive textiles, clothing, and other light manufactures had limited growth potential owing to encroachment by other developing countries. The country's comparative advantage was shifting increasingly towards such items as machinery, electronics, steel, shipbuilding, and motor cars from the early 1970s. Accordingly, South Korea began to put more gross fixed investment in these areas; thus, the share of these manufactures was boosted from 13% of total manufactured output in 1975 to nearly 21% in 1981. However, the economy had to face both internal and external shocks during the fourth Five Year Plan (1977–81) period. The internal shock accompanied the political disruption following President Park Chung-Hee's assassination in October 1979 and a disastrous harvest in 1980. The impact was amplified by the turbulent transition and martial law enforcement in 1980.

The external shock, meanwhile, was related to the jump in oil prices following the second oil shock in early 1979, which was worsened by over-enthusiastic implementation of heavy industrial investment in South Korea. During the fourth Five-Year Plan period, these shocks, coupled with an allotment of about 80% of manufacturing investment to the heavy and chemical industries, contributed to a high rate of inflation, reaching a peak of 27% in 1980 when real GNP dropped for the first time since 1962.

The dramatic reversal of 1980 and a newly installed government under General Chun Doo-Hwan emphasized price stabilization above all economic options. Since stabilizing monetary expansion, South Korea's exports have begun to soar following the Korean *won* depreciation of more than 44.6% between 1980 and 1986. With the turning point of 1986, South Korea came to show a trade surplus reaching about $3130.6 million. Since then, it has enjoyed larger surpluses ($6261 million in 1987, $8886 million in 1988 and $912 million in 1989) for four years in a row. This remarkable trade surplus was largely attributed to the devaluation of the *won*, accompanied by domestic monetary stabilization, low oil prices, and low interest rates in the world markets during the 1982–6 period.

A devaluation (a decrease in the value of domestic currency in terms of foreign currency) can, theoretically and empirically, increase a country's trade balance provided it is not accompanied by a monetary overhang. A devaluation makes imports more expensive to domestic residents and hence decreases the real quantity of goods imported, and it makes exports less expensive to foreigners and hence increases the real quantity of goods exported. But this does not guarantee that the balance of trade will improve. If the elasticity of demand for imports is sufficiently less than 1 in absolute value, the effect of the increase in (domestic) money expenditure on imports on the balance of trade may not be offset by the increase in (domestic) money revenue from the increased sales of exports, particularly if the elasticity of demand for the exports is also small.

A simple statement of this criterion is the Marshall–Lerner condition, which says that devaluation will improve the trade balance if the sum of the elasticities of imports and exports is greater than unity. If, on the other hand, a weaker currency is brought about by a monetary expansion, then it usually has an ambiguous short-run effect on the balance of trade. Put in more conventional terms, the weaker (domestic) currency resulting from the monetary expansion induces a rise in exports, but it can also lead to a rise in imports, since domestic spending is increased by lower domestic real interest rates. Therefore, the sources of exchange rate shifts are important if they are to be effective. South Korea achieved price stabilization of the annual average wholesale inflation rate of less than 0.8% during the period 1982–7. Thus, the currency depreciation of about 44.6% between 1981 and 1986 without a doubt had an effect on the lagged improvement of the trade balance. Table 5.7 shows the trends of South Korea's trade growth, price index, and exchange rate from 1952 to the end of 1990.

A striking characteristic of the South Korean economy is the growing share of exports in GNP and the changing composition of its trade. Exports were worth less than $40 million in 1953 when the Korean War ended; they accounted for only 3% of the country's GNP at that time. This proportion steadily grew to about 14.5% in 1963–73, 25.8% in 1974–80, and 32.8% in 1981–9. In 1990, exports of goods reached $65 billion, and with services was equal to about 43% of GNP.

With this continuous rise in exports, South Korea has moved from being a primary commodity exporter to a net importer of commodities and exporter of manufactures. In the 1950s and early 1960s, most South Korean exports were primary commodities—mainly tungsten, iron ore, fish, raw silk, agar-agar, rice, coal, etc.

TABLE 5.7 South Korea's external transactions, 1950–1990

	Current balance[a] (US$ million)	Trade balance[a] (US$ million)	Exports[b] (US$ million)	Imports[b] (US$ million)	Wholesale price index 1985 = 100	Exchange rate (W/$)
1950	23.1	−54.6	—	—	—	2.50
1951	52.8	−113.9	—	—	—	6.00
1952	61.2	−186.4	27.7	214.2	0.95	6.00
1953	−67.8	−305.7	39.6	345.4	1.19	18.00
1954	−33.4	−218.2	24.2	243.3	1.53	18.00
1955	−36.3	−309.9	18.0	341.4	2.76	50.00
1956	−23.1	−354.8	24.6	386.1	3.63	50.00
1957	−2.6	−370.9	22.2	442.2	4.22	50.00
1958	37.7	−326.4	16.5	378.2	3.96	50.00
1959	16.4	−253.6	19.8	303.8	4.07	50.00
1960	13.4	−272.5	32.8	343.5	4.50	65.00
1961	33.1	−242.2	40.9	316.1	5.10	130.00
1962	−55.5	−335.3	54.8	421.8	5.58	130.00
1963	−143.3	−410.2	86.8	560.3	6.73	130.00
1964	−26.1	−244.9	119.1	404.4	9.06	256.02
1965	9.1	−240.3	175.1	463.4	9.97	272.06
1966	−103.4	−429.5	250.3	716.4	10.85	271.46
1967	−191.9	−574.2	320.2	996.2	11.54	274.60
1968	−440.3	−835.7	455.4	1 462.9	12.48	281.50
1969	−548.6	−991.7	622.5	1 823.6	13.32	304.50

1970	−622.5	835.2	1 984.0	14.54	316.70
1971	−847.5	1 067.6	2 394.3	15.80	373.20
1972	−371.2	1 621.1	2 522.0	18.00	398.90
1973	−308.8	3 225.0	4 240.3	19.25	397.50
1974	−2 022.7	4 460.4	6 851.8	27.36	484.00
1975	−1 886.9	5 081.0	7 274.4	34.61	484.00
1976	−313.6	7 715.3	8 773.6	38.80	484.00
1977	12.3	10 046.5	10 810.5	42.30	484.00
1978	−1 085.2	12 710.6	14 971.9	47.25	484.00
1979	−4 151.1	15 055.5	20 338.6	56.09	484.00
1980	−5 320.7	17 504.9	22 291.7	77.96	659.90
1981	−4 646.0	21 253.8	26 131.4	93.86	700.50
1982	−2 649.6	21 853.4	24 250.8	98.24	748.80
1983	−1 606.0	24 445.1	26 192.2	98.40	795.50
1984	−1 372.6	29 244.9	30 631.4	99.12	827.40
1985	−887.4	30 283.1	31 135.7	100.0	890.20
1986	4 617.0	34 714.5	31 583.9	98.5	861.40
1987	9 853.9	47 280.9	41 019.8	99.0	792.30
1988	14 161.0	60 696.0	51 811.0	101.7	684.10
1989	5 055.0	62 377.0	61 465.0	103.2	679.60
1990	−2 179.4	65 016.0	69 844.0	107.5	716.40

[a] f.o.b. basis.
[b] Customs clearance basis.
Sources: Bank of Korea, *Principal Economic and Social Indicators, 1945–1988* (October 1988); and *Principal Economic Indicators* (each issue).

TABLE 5.8 South Korea's trade dependency ratio, 1970–1990 (US$ billion)

	GNP (A)	Exports[a] and imports (B)	Exports (C)[a]	B/A (%)	C/A (%)
1970	8.1	2.8	0.8	34.6	9.9
1971	9.5	3.5	1.1	36.8	11.6
1972	10.7	4.1	1.6	38.3	15.0
1973	13.5	7.4	3.2	54.8	23.7
1974	18.8	11.4	4.5	60.6	23.9
1975	20.9	12.4	5.1	59.3	24.4
1976	28.7	16.5	7.7	57.5	26.8
1977	36.8	20.8	10.0	56.5	27.2
1978	51.5	27.7	12.7	53.8	24.7
1979	61.5	35.5	15.1	57.7	24.6
1980	60.5	39.8	17.5	65.8	28.9
1981	66.8	47.4	21.3	71.0	31.9
1982	71.3	46.2	21.9	64.8	30.7
1983	79.5	50.6	24.4	63.6	30.7
1984	87.0	59.8	29.2	68.7	33.6
1985	89.7	61.4	30.3	68.5	33.8
1986	102.7	66.3	34.7	64.6	33.8
1987	128.9	88.3	47.3	68.5	36.7
1988	172.8	112.5	60.7	65.1	35.1
1989	211.2	123.8	62.4	58.6	29.5
1990	237.9	134.9	65.0	56.7	27.3

[a] Customs clearance basis.

Sources: Bank of Korea; Office of Customs and Tariffs (Seoul).

Manufactured exports made up only a small fraction of total exports until 1962, but thereafter expanded more rapidly than the exports of primary commodities. By 1973, such manufactured goods as clothing, electrical machinery, plywood, textile fabrics, iron and steel sheets, footwear, and wigs became the major components of South Korean exports. Manufactured exports rose from only 27% of total exports in 1962 to 88% in 1973 and to 94.5% by 1989. In the 1980s, this proportion remained at around 95%, of which heavy industrial products occupied more than a half. Light industrial products accounted for about 40% of total exports.

In imports, raw materials, fuel, and chemicals together accounted for just about half the value of total imports in the 1950s. But as resource-poor and densely populated South Korea continues to grow, its dependence on imports both for the production of exports and for domestic consumption also increases rapidly. The import of raw materials alone accounted for 38.1% of all imports in 1983, and rose to 45.6% by 1989, reflecting in part the import content of South Korea's exports. Capital goods imports increased from 29.8% in 1983 to 36.4% in 1989, while chemicals remained at around 19.0% of the total value of imports. The proportion of crude oil imports dropped from 21.3% in 1983 to 8.0% in 1989, reflecting the downhill trend of international crude oil prices until the outbreak of the Gulf crisis in August 1990.

Tables 5.9 (*a*) and (*b*) present the destination of South Korea's exports and the sources of its imports for selected years between 1953 and 1991. Throughout the period, more than half of South Korean exports and imports have been to and from two rich countries: the United States and Japan. In 1954 about 90% of South Korea's exports went to these two countries; then the share declined to about 60% of total exports in the first half of the 1960s before again increasing to roughly 70% in the late 1960s and early 1970s. The temporary drop in South Korea's exports to the United States in the first half of the 1960s apparently reflected the political disharmony between the Park Chung-Hee military government and the United States.

Nevertheless, the United States has remained South Korea's single largest trade partner. In fact, US economic and military aid and support have unquestionably played a large role in creating the environment that has made South Korea's rapid growth possible. The United States has so far provided approximately $13.6 billion in economic and military assistance to Seoul, not to mention the US–ROK Mutual Defence Treaty, still in effect at the end of 1990. The bilateral trade between the United States and South Korea increased from a mere $150 million in 1960 to $36.5 billion in 1989. As a result, South Korea has become Washington's seventh largest trading partner. It goes without saying that the United States has provided the largest market for South Korean exports, although Seoul's trade surplus with the United States began to drop in 1989, to $4.7 billion, and again to $0.9 billion for the first half of 1990 from its peak of $9.6 billion in 1987. The US share of

TABLE 5.9(*a*) South Korea's exports, by major country, 1955–1991[a] (US\$ million[b])

Country	1955	1965	1975	1985	1989	1990	1991
USA	3.79	61.7	1 536.3	10 754.0	20 639.0	19 360.0	18 559.3
	(41.9)	(35.2)	(30.2)	(35.5)	(33.1)	(29.9)	(25.9)
Japan	3.53	44.0	1 292.9	4 543.0	13 457.0	12 637.9	12 355.8
	(39.0)	(25.1)	(25.4)	(15.0)	(21.6)	(19.4)	(17.2)
Hong Kong	1.01	10.8	182.0	1 566.0	3 375.0	3 779.9	4 769.0
	(11.2)	(6.2)	(3.6)	(5.2)	(5.4)	(5.8)	(6.6)
West Germany[c]	0.14	3.2	312.3	979.0	2 137.0	2 849.2	3 192.4
	(1.5)	(1.8)	(6.1)	(3.2)	(3.4)	(4.3)	(4.4)
UK	0.06	3.6	161.8	913.0	1 861.0	1 750.4	1 767.5
	(0.7)	(2.1)	(3.2)	(3.0)	(3.0)	(2.7)	(2.5)
Canada	0.01	2.5	197.3	1 229.0	1 882.0	1 730.8	1 672.9
	(0.1)	(1.4)	(3.9)	(4.1)	(3.0)	(2.7)	(2.3)
Taiwan	—	1.9	64.2	196.0	1 308.0	1 248.6	1 609.0
		(1.1)	(1.3)	(0.6)	(2.1)	(1.9)	(2.2)
Australia	—			369.0	1 005.0	956.0	990.0
				(1.2)	(1.6)	(1.5)	(1.3)
France	—	0.5	40.1	316.0	894.0	1 118.9	1 127.9
		(0.3)	(0.8)	(1.0)	(1.4)	(1.7)	(1.6)
Saudi Arabia	—			969.0	815.0	739.7	980.3
				(3.2)	(1.3)	(1.1)	(1.3)
Others	0.50	46.9	1 294.1	8 449.0	15 004.0	18 844.3	24 846.0
	(5.5)	(26.8)	(25.5)	(27.9)	(24.1)	(29.0)	(34.7)
Total	9.04	175.1	5 081.0	30 283.0	62 377.0	65 015.7	71 870.0
	(100.0)	(100.0)	(100.0)	(100.0)	(100.0)	(100.0)	(100.0)

[a] Customs clearance basis.
[b] Figures in parentheses are percentages of the total.
[c] 1991 figures still pertain to what was formerly West Germany.

Country	1955	1965	1975	1985	1989	1990	1991
Japan	7.07 (14.7)	166.6 (36.0)	2 433.6 (33.5)	7 560.0 (24.3)	17 449.0 (28.4)	18 573.9 (26.6)	21 120.2 (25.9)
USA	16.85 (34.9)	182.3 (39.3)	1 881.1 (25.9)	6 489.0 (20.8)	15 911.0 (25.9)	16 942.5 (24.3)	18 894.4 (23.2)
West Germany[c]	2.17 (4.5)	16.1 (3.5)	192.7 (2.6)	979.0 (3.1)	2 624.0 (4.3)	3 283.6 (4.7)	3 698.3 (4.5)
Australia	—	—	—	1 116.0 (3.6)	2 243.0 (3.6)	2 589.1 (3.7)	3 009.4 (3.7)
Canada	0.10 (0.2)	1.60 (0.4)	150.2 (2.1)	630.0 (2.0)	1 680.0 (2.7)	1 465.4 (2.1)	1 906.9 (2.3)
Malaysia	0.41 (0.8)	—	—	1 234.0 (4.0)	1 503.0 (2.4)	1 586.0 (2.3)	1 869.0 (2.2)
Taiwan	2.34 (4.8)	10.5 (2.3)	164.4 (2.3)	333.0 (1.1)	1 328.0 (2.2)	1 451.9 (2.0)	1 514.7 (1.9)
Indonesia	—	—	146.8 (2.0)	669.0 (2.1)	1 135.0 (1.8)	1 600.3 (2.2)	2 051.8 (2.5)
Saudi Arabia	—	—	—	640.0 (2.1)	1 042.0 (1.7)	1 724.9 (2.5)	3 268.6 (4.1)
UK	1.97 (4.1)	1.2 (0.3)	123.0 (1.7)	653.0 (2.1)	923.0 (1.5)	1 226.1 (1.8)	1 558.1 (1.9)
Others	17.34 (35.9)	85.1 (18.4)	2 182.6 (30.0)	10 833.0 (34.8)	15 627.0 (25.4)	19 400.0 (27.8)	22 632.7 (27.8)
Total	48.25 (100.0)	463.4 (100.0)	7 274.4 (100.0)	31 136.0 (100.0)	61 465.0 (100.0)	69 843.7 (100.0)	81 524.9 (100.0)

[a] Customs clearance basis.

[b] Figures in parentheses are percentages of total.

[c] 1991 figures still pertain to what was formerly West Germany.

Sources: South Korea's Foreign Trade Association; Bank of Korea.

South Korea's total exports was about 36.4% in the 1970s and 33.3% in the 1980s. US exports accounted for 25.0% of South Korean total imports in the 1970s, then dropped slightly to 22.8% in the 1980s.

However, South Korea's continuous flooding of US markets with cheaper goods and its trade surplus with the United States since 1982 have led to recent pressure from Americans to 'level the playing field'. This reflects in part the changing state of US–Korea relations as well as the change in global economic and political environments.

It must be pointed out in passing that South Korea's outward-oriented growth strategy has not represented a *laissez-faire* approach until recently. Intervention in the form of trade restrictions such as tariffs and licensing, subsidies, tax incentives, and credit allocation have been constant and have enabled Korean firms to maintain an active contact with the world economy on both the import and export side.[19] The credit system has channelled financial resources at subsidized rates to preferred activities. The tax system has provided an exemption from import duties for exports, favourable tax rates on profits and incomes from goods exported, and direct cash subsidies for businesses producing goods for export.

In this way, South Korea has become one of the 'little dragons' in Pacific Asia, inviting its trading partners to observe carefully its idiosyncratic practices in foreign trade. Since 1989, however, the wind has changed for the South Korean economy, with endless labour disputes in domestic markets and ever-mounting pressures from external partners regarding future export markets. South Korea's short-lived trade surplus began to shrink, from $11 445 million in 1988 to only $4597 million in 1989, and to a $2004 million *deficit* at the end of 1990.

However, it is noteworthy that South Korea's trade with communist bloc nations dramatically increased, by 23.7% to $3.77 billion in the first nine months of 1990 from $3.05 billion in the same period of 1989. This represents 3.9% of the total trade of South Korea, up 0.5 percentage points from 1989, compared with 0.1% in 1984. South Korea imported goods to the value of $1.99 billion and exported $1.78 billion worth of manufactured goods, running up a trade deficit of $210 million with the communist bloc countries (see Table 5.10).

TABLE 5.10 South Korea's trade with CPEs, 1986–1988 (US$ million)

Country	Exports			Imports		
	1986	1987	1988	1986	1987	1988
USSR	65.00	67.00	112.00	68.00	97.00	178.00
Poland	18.99	20.90	27.12	5.42	5.52	15.25
East Germany	21.24	37.57	28.87	3.96	3.04	7.62
Hungary	7.32	14.99	20.69	2.52	7.70	11.44
Yugoslavia	9.67	15.98	16.63	1.06	2.18	17.32
Bulgaria	1.09	1.28	5.54	2.59	2.12	3.62
Czechoslovakia	7.09	8.56	23.37	3.31	5.37	11.18
Roumania	2.15	3.08	2.76	0.78	18.28	22.89
Total	132.54	169.35	237.57	87.64	143.20	267.32

Sources: KOTRA; OECD, *Economic Surveys 1987/88* (Paris: OECD, 1988); UN Economic Commission for Europe, *Economic Survey of Europe in 1988–89*.

Trade with these countries grew rapidly following the 1988 Olympics in Seoul, perhaps as a result of the South's 'northern' policy of establishing diplomatic ties and expanding economic co-operation with communist countries. China, where Seoul had just set up a trade office in October 1990, took the lead with $3.82 billion in two-way trade, followed by the Soviet Union, which normalized diplomatic relations in September 1990, with $889 million by the end of that year. South Korea–China trade first started indirectly in the early 1970s and reached $3821 million in 1990 (see Table 5.11). The sharp surge in bilateral trade is noteworthy, since it took place amidst the sharp drop in China's trading volume with other countries owing to the Tiananmen Square massacre on 4 May 1989. China was the fourth largest trading partner of South Korea, while Seoul was Beijing's eighth, at the end of 1989. Out of a total of $2347 million of Seoul's imports value with centrally planned economies, China accounted for 68% with $1589 million in 1989.[20] South Korea expects that the establishment of trade offices in Seoul and Beijing will escalate two-way trade in time to come.

Trade with Eastern European countries in the first nine months of 1990 was $517 million, up 101.2 percentage points from the

TABLE 5.11 South Korea's trade with China, 1979–1990 (US$ million)

	Exports	Imports	Trade balance	Total trade
1979	4	15	−11	19
1980	115	73	42	188
1981	205	148	. 57	353
1982	48	81	−33	129
1983	51	83	−32	134
1984	229	233	−4	462
1985	683	607	76	1 290
1986	714	680	34	1 394
1987	813	673	140	1 486
1988	1 700	1 387	313	3 087
1989	1 554	1 589	−35	3 143
1990	1 553	2 268	−715	3 821

Sources: KOTRA estimates; Ministry of Commerce and Industry (Seoul).

previous year. South Korea's main exports to these communist countries are TV sets and parts, refrigerators, car batteries, textiles, videotape recorders, footwear, tennis rackets, silk blouses, and soaps, while its principal imports are cement, crude oil, marine products, raw cotton, and machine tools.

Business leaders from South Korea and these CPE countries appear to recognize the need for further promoting co-operation in trade, joint ventures, and technology transfer between one another. Above all, the now-changing Eastern bloc countries appear to have a mounting interest in learning how this resource-poor country has attained its economic miracle within the competitive economic system. Often the very central elements embedded in the South Korean competitive paradigm draw the attention of Eastern bloc economists who are seeking to transform their economies into market-based systems. Paradoxically, perhaps the most peculiar lesson that can be learned from the South Korean experience is not the advantages of free enterprise, but the effective role of a strong state. However, caution must be in order, because it is doubtful if the South Korean model would be suitable for different cultural and political environments.

5.2.2 *External Resources and Economic Co-operation*

One important factor that cannot be underestimated is the foreign resources South Korea has received both for its survival and for economic development projects. Indeed, foreign resources have played a most important role in the economic development of South Korea, as it did in many of today's economically advanced countries.[21] The ratio of capital inflow to gross domestic capital formation in South Korea was about 45% in the first Five Year Plan (FYP) (1962–6), 42% in the second FYP (1967–71), 6.6% in the third FYP (1972–6), and 32% in the fourth FYP (1977–81). The ratio, however, decreased dramatically to be a negative number from the sixth Five Year Plan (1982–6), both because a higher domestic saving ratio exceeded the gross investment ratio and because of its improvement in the balance of trade.

In the early stage of economic development, foreign resources are usually mobilized to replenish the dearth of domestic savings in a developing country. Generally, the difference between planned investment (I) and planned savings (S) is taken as an indication of the foreign resources (F) that are necessary to attain a target rate of economic growth. That is,

$$F = I - S.$$

Indeed, in many development planning models, a target level of investment is specified to achieve a certain rate of growth of income and then an estimate of planned savings is made. If the planned investment exceeds planned savings, the country seeks to bridge the gap with imported foreign capital.

In this 'savings gap' analysis, notice the implicit assumption that all foreign resources will be used for domestic investment. This need not always be true, and it is possible to see the coexistence of both a 'savings gap' and a 'foreign-exchange' or 'trade' or 'bottleneck gap', where such a gap (T) is given by the difference between imports (M) and exports (X). Thus,

$$T = M - X.$$

The equilibrium relationship between the 'savings gap' and the 'trade gap' can be expressed as

$$I - S = M - X.$$

Notice that these two gaps are equal *ex post* because of the method of national accounting. It is contended that, where the trade gap dominates over the savings gap, a supply of foreign resources could have a positive effect on growth.[22]

Of course, the effects of foreign resources on domestic economic growth and development must depend on the host country's ability to absorb such resources and to use them in an effective way. Such 'effective' use of foreign resources can be measured in part by positive reasonable rates of return of capital on total investment, which would depend upon the supply of managerial skills as well as the level of average and marginal rates of saving in the host country. It also depends on the composition of foreign resources between 'grants', which need not be repaid by the recipient countries and do not carry any interest charges, and 'loans and credit', which must be repaid with interest. The marginal rate of return to capital was very negligible until the late 1950s, and then it turned positive by a factor of about 4–6% annually in the 1960s and 1970s. Capital input has accounted for about 13.4% in the 1963–72 period and 26.3% in 1972–82 of the sources of growth in South Korea. On the other hand, labour input and total factor productivity accounted for about 37.8% and 48.8% respectively in 1963–72, and for 43.8% and 30.0% in 1972–82.[23]

South Korea's economy was characterized first and foremost by its heavy dependence on US aid and grants between 1945 and 1961. Foreign aid and grants financed nearly 70% of South Korea's imports and investments. Counterpart funds generated by sales of agricultural products supplied under US Public Law 480 accounted for half of the South Korean budget at this time.[24]

Although the overall rate of economic growth during this period was discouraging, changes were taking place in the economy that would improve the prospects for 'take-off' when a growth-oriented government came to power. During the early 1960s, large amounts of aid poured into the country to help the poverty-stricken South sustain 'survival', repair the infrastructure, and rebuild the war-damaged nation. As shown in Table 5.12, South Korea had received a total of $4728.6 million in economic aid (and $8871.8 million in military assistance) by 1982, of which $3559.3 million of economic aid, 75%, was received by 1963. Starting in 1962, South Korea gradually shifted from an aid and grants-recipient country to a borrowing one on the basis of both public and commercial loans

TABLE 5.12 Foreign capital received by South Korea, 1945–1991
(US$ million)

	Aid	Public loans	Commercial loans	Direct investment in Korea
1945	4.4	—	—	—
1946	49.5	—	—	—
1947	175.4	—	—	—
1948	179.6	—	—	—
1949	116.5	—	—	—
1950	58.7	—	—	—
1951	106.5	—	—	—
1952	161.3	—	—	—
1953	194.2	—	—	—
1954	153.9	—	—	—
1955	236.7	—	—	—
1956	326.7	—	—	—
1957	382.9	—	—	—
1958	321.3	—	—	—
1959	222.2	2.2	—	—
1960	256.1	2.2	—	—
1961	206.9	1.2	—	—
1962	200.0	1.8	2.6	0.6
1963	207.5	40.3	20.5	4.8
1964	141.0	14.8	13.2	0.7
1965	134.6	5.0	41.5	5.6
1966	122.0	73.1	118.7	13.4
1967	134.5	105.6	124.0	7.6
1968	120.6	70.2	268.4	19.2
1969	103.9	213.8	410.6	15.9
1970	85.1	209.9	366.7	66.1
1971	64.0	303.4	345.2	42.9
1972	50.6	324.5	326.4	78.8
1973	35.1	368.5	344.4	143.3
1974	41.4	316.7	616.0	124.1
1975	48.8	482.1	804.5	61.6
1976	14.4	710.7	842.5	85.5
1977	8.0	608.0	1 260.0	104.4
1978	10.6	817.7	1 929.8	100.5
1979	11.8	1 085.6	1 621.8	126.0
1980	17.1	1 518.3	1 415.8	96.2
1981	13.3	1 704.5	1 242.5	105.4

TABLE 5.12 (*cont.*)

	Aid	Public loans	Commercial loans	Direct investment in Korea
1982	12.5	1 877.3	918.4	100.6
1983	—	1 493.9	973.2	101.4
1984	—	1 424.4	858.4	170.7
1985	—	1 023.8	964.1	250.3
1986	—	880.1	1 619.9	477.5
1987	—	1 109.0	1 558.3	624.8
1988[a]	—	891.0	988.0	894.1
1989[a]	—	475.3	859.7	813.1
1990[a]	—	418.0	30.0	89.5
1991[a]	—	429.0	—	117.5

[a] Foreign investments in Korea have decreased since 1988; a growing number of foreign business concerns are returning home, apparently affected by high labour costs and the worsening business climate.

Sources: Ministry of Finance, *Finance Statistics* (monthly); Economic Planning Board, *Major Statistics of Korean Economy* (1965, 1989, 1991).

as well as direct foreign investment, reflecting the country's better economic position.

There have been arguments about the positive and negative aspects of the contribution of US aid to South Korea during 1945–60. The Harvard–Korean Development Institute study concludes that the high level of US aid made all the difference between the 1.5% annual real increase in per capita income achieved and no growth at all. 'Without this growth, the economic condition of the population would have remained desperate, political cohesion would have deteriorated, and the foundations for subsequent high growth would not have been forged.'[25]

Another contribution of the earlier US aid was that it made it possible for South Korea to contract virtually no debt obligations in the start-up period of its industrialization, unlike most Third World countries. In the absence of any such foreign debt obligations in the early 1960s, South Korea could favourably import from credit-offering countries the foreign savings needed to ignite and ensure its economic development schemes.

The (opposing) view sees US aid and grants as merely bringing about South Korea's chronic dependency on non-productive transfer income, while hampering the ability to accumulate capital within the economy. In addition, foreign aid had been closely linked with the growth of rampant corruption of South Korean bureaucrats, thus creating a group of selfish monopolistic capitalists whose actions only led to weak economic conditions. Critics also argue that the aid under US Public Law 480 in particular resulted in the gradual ruin of the foundations for farm production in South Korea.[26] The agricultural sector has decreased tremendously, from 47.2% in 1956 to about 10.2% by 1989 in the composition of GDP. The net foreign debt outstanding mounted to a peak of $46.7 billion at the end of 1985. It appears as if the pessimists' view of this was correct.

Nevertheless, South Korea's industrialization and its remarkable economic growth is not fictitious. As the South Korean trade balance improves, the country is now recognized as one of the 'miracle' economies which are so successful in economic development with foreign savings. South Korea's net foreign debts dropped to $3.0 billion at the end of 1989, while its assets abroad increased from $11.2 billion in 1985 to $26.4 billion in 1989, a 135.7% rise within a five-year period. In 1988 its debt service ratio dropped to the single-digit level (see Table 5.13). Today no one can fear that South Korea is likely to face the risk of default, although it still had about $29.4 billion of total foreign debts outstanding in 1989.

In recent years, however, foreign investors have deserted South Korea in increasing numbers, apparently affected by the worsening domestic business climate since 1988. The decline in foreign investment is largely ascribable to recent rising labour costs in labour-intensive industries. And there is also some worry about the increasingly uncompetitive South Korean agricultural sector accompanied by the dilemma of rural underemployment facing about 20% of total population in the South. Furthermore, the ongoing bilateral trade friction between the US and South Korea over access to agricultural markets in the South raises the issue of adjustment in the structure of agriculture. Certainly South Korea faces serious problems of adjustment in agriculture, primarily because for the past several decades it has neglected to provide an off-the-farm solution to decreasing farm production. The opening

TABLE 5.13 South Korea's outstanding external debt, 1962–1989 (US$ million)

	External debt outstanding (A)	Foreign asset holdings (B)	Net external debt (A – B)	Debt service ratio[a] (%)	Ratio of net[b] debt to GNP (%)
1962	166	—	166	—	7.2
1967	473	—	473	9.4	11.0
1972	3 590	—	3 590	20.0	33.6
1973	4 261	—	4 261	13.0	31.6
1974	5 939	—	5 939	11.0	31.6
1975	8 448	—	8 448	10.9	40.4
1976	10 524	—	10 524	9.7	36.7
1977	12 646	—	12 646	9.8	34.4
1978	14 866	—	14 866	11.4	28.9
1979	20 300	6 300	14 000	13.2	22.8
1980	27 200	7 500	19 600	12.2	32.4
1981	32 400	8 000	24 500	11.9	16.4
1982	37 100	8 800	28 300	13.0	20.2
1983	40 400	9 500	30 900	10.4	38.9
1984	43 100	10 100	32 900	10.6	37.8
1985	46 700	11 200	35 500	10.0	39.6
1986	44 500	12 000	32 500	11.2	31.6
1987	35 600	13 200	22 400	13.5	17.4
1988	31 200	23 900	7 300	8.5	4.2
1989	29 400	26 400	3 000	5.4[c]	1.4

[a] The ratio of the sum of principal and interest payments to the value of exports (customs clearance basis).
[b] For 1962–78, total external debts were used for calculation.
[c] Excluding interest payment. This is the ratio of principal payments to the value of exports (customs clearance basis).

Sources: Ministry of Finance, *Finance Statistics* (monthly); Bank of Korea, *Economic Statistics Yearbook* (each issue).

of agricultural markets and the consequent realignment of domestic prices towards world levels would seriously undermine South Korea's rural household incomes and exacerbate sensitive regional income inequalities, which are increasingly becoming a social problem.

South Korea is well aware that it has been predominantly on the receiving side of international co-operation, and that it has now to make a positive contribution as a true member of the family of nations. Feeling an obligation to return what it has received to other countries, it appears to be increasing its participation in international economic organizations. At GATT and the Uruguay Round talks, South Korea is supporting the removal of protectionist measures for the enhancement of free trade among nations. It also increased its funding to the International Monetary Fund (IMF), and is actively participating in the programmes for co-operation among developing countries of the UN Conference on Trade and Development (UNCTAD). South Korea was elected the vice-chair-country of the Economic and Social Commission for Asia and Pacific (ESCAP). In addition, it is promoting Asian-Pacific economic co-operation (APEC) on the principle of non-exclusivity of international relations; that is, countries should feel free to have extensive relations with countries outside the region.

Being still at a relatively early phase of economic take-off, South Korea can extend only a modest amount of economic assistance or investment abroad, no matter how eager it may be to do more. Although its resources are still very limited, it is sharing its development experiences with other developing countries wherever such experience is readily applicable and relevant.

Meanwhile, Seoul's 'northern' policy, initiated first by the South's President Roh Tae-Woo in his address on 7 July 1988, aims actively to increase cultural, trade, and economic co-operation with North Korea, as well as with Eastern European countries, China, Mongolia, and Russia. South Korea has already established diplomatic ties with most Eastern European countries, Mongolia, and Russia, beginning with Hungary in early 1989. Currently it has committed itself to providing Hungary with $250 million in bank loans over a four-year period from 1989. In addition, Seoul is now propelling joint development of resources in Siberia with the former Soviet Union, while it proposes construction of a gas pipeline linking South Korea and Russia via North Korea. The pipeline project is likely to be conducive to easing tension on the Korean peninsula, if a tripartite agreement is made.

5.3 FUTURE PROSPECTS

In the sphere of foreign trade, South Korea has broadly pursued an export-oriented strategy, while North Korea has called for a disproportionate emphasis on the domestic development of heavy industry based on its *chuche* ideology. The two different economic systems and ideologies have resulted in different development and trade patterns as well as different incentives facing the two populations. The external trade value of South Korea was merely $376.3 million in 1960, which slightly exceeded that of North Korea of $320 million. Over the past three decades, however, South Korea's trade has become about 26 times larger than the size of North Korea's exports and imports in value. In 1989, the trade value of the South reached $123 842 million as contrasted to $4078 million of the North. Indeed, the difference in policy emphasis as well as in economic systems has led to a fundamental gap between North and South Korea's economies and development patterns, influencing such factors as preferences, technology, endowments, and the decision-making environment, which are considered to be the crucial exogenous variables in the analysis and comparison of the two economies.

Generally speaking, South Korea was not bound by economic ideologies that set objectives or priorities, in contrast to North Korea. Although until recently the South Korean government had a complete monopoly on all institutional credit and exercised great power in intervening or influencing industries or enterprise areas in accordance with its perceived national interests, the basic decision-making on industrial production or trade came from the private sector. Market-oriented competition encouraged efficiency, led to a diversification of products, and fostered a growing class of able entrepreneurs and business managers. These factors have helped South Korea adjust flexibly, in line with its export-oriented strategy, rooted in the notion of the Ricardian theory of changing comparative advantages in trade. The concentration on labour-intensive manufactures for export in the early stage of its development shifted to capital-intensive industries such as shipbuilding, steel, and the chemical industries from the early 1970s. Such a shift, initiated largely by private enterprise and government support, was inevitable and timely, as the steady increase in the real wages of South Korean workers made their products less competitive with goods produced by countries with lower wage levels. The government policy of export-led growth has worked together with profit-motivated enterprises which have constantly sought to digest, adopt, and adapt to new environments and technology.

In contrast, the North Korean communist system of government has established a high degree of centralization and planning. It is largely characterized by a pronounced bias towards investment in heavy industry and defence and some showcase monuments of 'social construction' including the Tower of the Chuche Idea, the enormous 150 000-seat May First Stadium, and the 3000-room (empty) high-rise Ryugyong Hotel.[27] Its emphasis on self-reliance hindered its expansion of foreign trade as well as its access to modern technology.

Between 1985 and 1989, North Korea imported only $9728.1 million worth of goods, while South Korea imported almost $217.0 billion. The early attempt to remedy such a trade-lag ended in failure, not only because of the country's foreign debt default crisis in the mid-1970s, but also because of President Kim Il-Sung's underlying unwillingness to become less autarkic by expanding foreign trade for fear of contamination by outside influences. However, the situation is changing rapidly with time. As North Korea's two close trading partners, Moscow and Beijing, established official relations with Seoul in the autumn of 1990, Pyongyang is being forced to retract its reclusive *chuche* principle and to seek to make new friends in the West.

South Korea is keenly aware of the North's underlying desire to embrace a more open policy, and is seeking positively to establish interrelations with the North through a series of high-level talks. However, even if there are commonalities in both North and South based on a shared culture and language, the two parts of the country have gone along different roads in both politics and economics for 45 years now.

North Korea, in a sense, chose politics over economics. It pushed the *chuche* ideology of autonomy effectively, isolating itself from much of the world. South Korea pursued a policy of external orientation, keeping economics largely separate from politics. It is remarkable that South Korea's foreign dependency policy has proved somewhat more effective than North Korea's self-sufficiency, at least in recent years. In the South, export-led growth has resulted in political liberalization and the destruction of the autocratic regimes that gave birth to rapid development; but in the North, *chuche* appears to have produced neither political freedom nor economic innovation so far.[28]

While the two Koreas both avow the desire for national unification the important question is how to integrate the North's communist system with the South's market-oriented economy in a transition such as the two Germanys are undertaking in their unification process.

NOTES TO CHAPTER 5

1. Charles Wolf, Jr, 'A Theory of Nonmarket Failure: Framework for Implementation Analysis', *Journal of Law and Economics*, 22 (April 1979), 107–39.
2. David I. Steinberg, *The Republic of Korea: Economic Transformation and Social Change* (Boulder, Colo.: Westview Press, 1989), 159.
3. See *Economic Dictionary* (Pyongyang: Social Science Publishing Co., 1970), 690.
4. *The Europa World Yearbook, 1989*, ii. 1559.
5. Ibid. 1560.
6. The trade dependency ratio was calculated by using North Korea's dollar GNP which was estimated with official exchange rates.
7. *EIU Country Report*, no. 1 (1988), 40.
8. KCNA report, 26 October 1990.
9. According to the Soviet TASS news agency reports on 22 September 1990, North Korea is showing signs of change, though not such conspicuous signs as in Eastern European and other ex-communist countries (FBIS-EAS-90–185, 24 September 1990, 24). It was also learned on 28 September 1990 that North Korea, while denouncing the Soviet Union's position on setting up diplomatic relations with South Korea, was eagerly seeking to improve relations with Japan and the USA. Such a drastic change in attitude, coupled with its official proposal to Japan's former Deputy Prime Minister Shin Kanemaru on improving relations with Japan in November 1990, is noteworthy because it indicates changes in the North-East Asian situation (*Seoul Sinmun* (in Korean), Seoul, 29 September 1990, 1). Regarding North Korea's position on reform and opening, read 'The true nature of "reform" and "opening" theory', DPRK *Information Bulletin* (Pyongyang), December 1989 (issued by the Secretariat of the Committee for Peaceful Reunification of Fatherland), 17–22.
10. Pyongyang KCNA (in English), 1038 GMT, 14 February 1990.
11. 'Minju Chosen [Free Korea] on Joint Venture Law', *The People's Korea* (Pyongyang), 8 May 1985.
12. Hy-Sang Lee, 'North Korea's Closed Economy: The Hidden Opening', *Asian Survey*, (University of California at Berkeley), (December 1988), 1272.
13. The Economist Intelligence Unit, *Country Report: China and North Korea*, no. 2 (1985), 37–8. See also *The Korea Times* (Chicago edn.), 19 November 1985.
14. FBIS-EAS-90–205, 25 October 1990, 33.
15. KCNA (Pyongyang), 7 October 1990, GMT 0831. See FBIS-EAS-90–208, 26 October 1990, 19.
16. KCNA (Pyongyang), 28 October 1990, GMT 0837. See FBIS-EAS-90–211, 31 October 1990, 24.
17. North Korea's Joint Venture Law and the establishment of an economic special zone are very similar to China's open-door policy in the early 1980s. To establish close links with the outside world, China promulgated a Joint Venture Law and tried to court foreign direct investment and technology. It also established four special economic zones, with wide-ranging tax concessions and other incentives to promote exports, between 1978 and 1982.
18. For the amounts of foreign assistance received by South Korea during 1945–61, see nn. 44 and 48 of Ch. 2.

19. South Korea increased its overall import liberalization ratio from 68.6% in 1980 to 96.4% in 1990, but its import liberalization ratio of agricultural products was scheduled to rise from 80.7% in 1990 to 84.9% by the end of 1991. It also reduced the overall average tariff rate from 23.7% in 1983 to 11.4% in 1990, while reducing tariffs on manufactures from 22.6% to 9.7% in the same period. The average tariff rate is scheduled to drop further, to 7.9% by 1993.

    It must be pointed out that regions such as Hong Kong and Singapore have pursued an export-oriented growth strategy while maintaining a free-trade regime almost from the beginning. But South Korea, Taiwan, and Japan pursued the export promotion strategy while maintaining a fairly protectionist import regime for a long period of time. Japan eliminated its quantitative restriction of imports by the early 1960s; the nominal import liberalization ratio expanded from less than 70% in 1960 to about 93% in 1964 and to 97% by 1976. Taiwan eliminated the quantitative restriction of imports by the early 1970s. However, imports have been very much restricted until recently in both Japan and Taiwan through special laws and other invisible unofficial means. South Korea eliminated the bulk of its tariff protection and quantitative restriction during the period 1984–8, and achieved a substantial import liberalization during the latter half of the 1980s.

20. *The Korea Herald* (Seoul), 21 October 1990, 6.

21. For example, between 1870 and 1914 the ratio of capital inflow to gross domestic capital formation was about 40% in Canada. This ratio for Australia was about 37% between 1861 and 1900, and for Norway it was 29% between 1885 and 1914, and 31% between 1920 and 1929 (see Everett E. Hagan, *The Economics of Development*, 2nd edn. (Homewood, Ill.: Richard D. Irwin, 1975)).

22. Hollis B. Chenery and A. Strout, 'Foreign Assistance and Economic Development', *American Economic Review*, 56 (1966), 680–733. Also see Ronald I. McKinnon, 'Foreign Exchange Constraints in Economic Development', *Economic Journal*, 74 (1964), 388–409.

23. Rudiger Dornbush, 'From Stabilization to Growth', paper presented at World Bank Annual Conference on Development Economics, Washington, DC, 26–27 April 1990, 16.

24. Edward S. Mason, Mahn-Je Kim, Dwight H. Perkins, Kwang-Suk Kim, and David C. Cole, *The Economic and Social Modernization of the Republic of Korea* (Cambridge, Mass.: Harvard University Press, 1980), 93–4, 103.

25. Ibid. 204.

26. Hyun-Che Park, *Korean Economy and Agriculture* (in Korean) (Seoul: Gachi Publication Co., 1983), 70–81, 83–101.

27. Pico Iyer, 'North Korea: In the Land of the Single Tune', *Time Magazine* (26 November 1990), 49–50.

28. Steinburg, *Republic of Korea*, 194.

6

North–South Relations Today

In the golden age of Asia
Korea was one of its lamp bearers
And that lamp is waiting to be lighted once again
For illumination in the East.

'To Korea', by Rabindranath Tagore

6.1 INTRODUCTION

Germany started the Second World War and was defeated and divided by the Allied Powers at the end of the war. But time passed, and an unpredicted wind that recently blew off the Berlin Wall led to reunification in October 1990. As Germans ratified reunification in their first free all-German election on 2 December 1990, many Koreans must have paused for a moment to consider the state of their own divided country and to reflect on the ironies of history. Korea too must now find its own way.

Korea remains an unsettled spot, the result of the growing Cold War confrontation of the early post-war years between the United States and the Soviet Union. With the recent dwindling of the Cold War, Koreans naturally feel a great uncertainty about the future of their split country. Every Korean must have had mixed feelings of expectation and apprehension about the first current North–South high-level contacts that began in September 1990 with a renewal of dialogue. Above all, the two Koreas officially became full members of the United Nations on 17 September 1991. The entry of both countries should be conducive to reducing tension on the Korean peninsula and promoting inter-Korean reconciliation, although there are still many unforeseen circumstances within and around the divided country.

Now is surely a time of change and uncertainty for people the world over. With the man-made walls falling down in Eastern Europe, the former communist bloc is engulfed in breathtaking changes. The world is currently witnessing transitional pains in the transformation of the Eastern bloc, the unification of Germany, and the collapse of the former Soviet Union and its domestic politics, economics, and foreign policy. The world is also sensing, from a distance, the Chinese struggle towards openness and economic and political reform and the internal conflicts in that process, and also North Korea's underlying turmoil, faced with changing external circumstances and the inevitability of a transfer of power from the cult figure Kim Il-Sung.

Above all, what is the core implication of the recent high-level talks between Seoul and Pyongyang in the light of German unification? Will the dialogue continue and eventually lead to historical breakthrough in the long-frozen relations, thus paving the way first for an adjustment of mutual differences and then for reconciliation and reunification?

There is, in fact, a wide gap in the systems themselves, as well as in the incumbent positions between the North and the South. The North has concentrated primarily on the solution to political and military problems, while the South placed greater importance on economic and humanitarian exchanges as a necessary step towards building mutual faith and trust. But of late the two sides seem to have come closer to accommodating each other's policies, which would appear to be a good sign for successful dialogue.

With such a hopeful interchange in progress, this chapter attempts to consider such practical matters as economic co-operation and trade between North and South Korea. For a thaw in inter-Korean relations could come sooner if the two sides work together to establish grounds for co-operation with the aim of removing the bones of contention and facilitating exchanges, beginning with trade based on mutual comparative advantage. This would serve to build a foundation for further co-operation, and could even pave the way for an eventual integration of the two countries.

6.2 THE STRUCTURE OF INDUSTRY AND TRADE IN NORTH AND SOUTH

An economic transaction occurs when the need arises between the parties concerned. The necessity of trade and the interaction between any two parties develops not only when potential traders have different comparative advantages in resources and basic commodities, but also when both parties are sure that they can gain from this mutual transaction. In other words, two parties participate in a transaction if both believe they will be better off by doing so. But economic motives are not necessarily the only reasons for deciding whether or not to engage in a physical transaction. Very often, political factors and selfish reasons of the ruling class exert an overriding influence against economic interaction, even though it might involve positive economic gains for both parties.

Indeed, economic trade could have benefited the two Koreas, which were endowed with different industrial structures and resources when the country was forced to divide. Instead, unfortunately, they were completely shut off from each other, without any economic interaction until recently.

North Korea promoted the *chuche* ideology of autonomy, effectively isolating itself from much of the world while investing too heavily in heavy and defence-related industry. The policies of its economic mentors appear to have resulted in a lower level of technology transfer, limited export opportunities, a shortage of foreign exchange, and relatively short supplies of light and consumer goods until recently.

In contrast, South Korea followed a policy of external orientation, allowing the economy to remain heavily dependent on foreign markets. Geographically, the South is blessed with rich soil and a mild climate, while the North is well endowed with mineral resources and industrial bases. Of course, the traditional economic structure has changed since 1950 as a result of the independent economic development of both sides. South Korea has now more industrial capacity than previously, while North Korea has added to her food-producing capacity. Table 6.1 compares the industrial structure between North and South for selected years. In the late 1980s, the importance of primary industry in total outputs was about 20% while secondary industry and other production accounted for 70% in North Korea, compared with about 10% and 33% respectively in the South.

Another aspect of the different economic systems is the relative share of the service (or non-productive) sector in the two economies. The non-productive sector in communist North Korea is very small, while in South Korea the service sector accounts for more than 50% of the economy.

TABLE 6.1 The industrial structures of North and South Korea, 1953–1990 (%)

	North Korea[a]					South Korea[b]		
	Agriculture, Fishery, and forestry		Industry	Other production sectors[c]	Non-production sectors[d]	Agriculture and fishery	Manufacturing and mining	Service sector
1953	41.6	(67.0)	30.7	20.9	6.8	47.3 (69.0)	10.0	42.7
1956	26.6	(40.0)	40.1	23.1	10.2	47.2 (64.9)	12.3	40.5
1960	19.3	(45.5)	62.3 (23.2)	13.6	4.8	38.0 (58.2)	20.5	41.5
1970	18.3	(41.4)	64.2 (27.6)	12.0	5.0	25.8 (50.4)	22.3 (14.3)	51.9 (35.3)
1975	10.0	(43.0)	66.0 (30.4)	12.0	12.0	25.0 (45.7)	27.5 (19.1)	47.6 (35.2)
1980	10.0	(45.9)	66.0	12.0	12.0	14.9 (34.0)	31.0 (22.5)	54.1 (43.5)
1983	—	(44.0)	—	—	—	13.6 (29.7)	31.0 (23.3)	55.4 (47.0)
1985	—	(40.5)	—	—	—	12.8 (24.9)	31.3 (24.4)	55.9 (50.6)
1987[e]	20.0	(38.0)	60.0 (37.0)	10.0 (12.0)	10.0 (13.0)	10.5 (21.9)	33.0 (28.1)	56.5 (50.0)
1988[e]	19.5	(38.0)	61.0 (37.5)	10.0 (12.5)	9.5 (12.0)	10.5 (20.7)	33.2 (28.5)	56.3 (50.9)
1989[e]	20.5	(37.5)	61.5 (38.0)	10.0 (12.5)	8.0 (12.0)	10.2 (17.3)	31.9 (28.2)	57.9 (52.3)
1990[f]	26.8	(—)	56.0 (—)	17.2 (—)	— (—)	9.1 (—)	44.8 (—)	46.1 (—)

[a] Composition of the GVSP. Figures in parentheses are the percentages of population in the relevant sectors.
[b] Composition of GNP. Figures in parentheses are the percentages of population in the relevant sectors.
[c] Basic construction, commerce, procurement, material, and technical supply.
[d] Transportation, communication, etc.
[e] Estimates for North Korea by this author.
[f] Estimates by South Korea's National Unification Board.

Sources: Chosen Chungang Nyungam (1965); *Rodong Sinmun* (27 January 1984); *Statistik des Auslandes: DVR Korea* (West Germany, 1982, 1986); National Unification Board (Seoul); Bank of Korea, *Economic Statistics Yearbook* (Seoul); and Economic Planning Board (Seoul).

TABLE 6.2 Total arable land and grain production in North and South Korea, selected years, 1949–1989

	North Korea			South Korea		
	1949	1967	1989	1949	1967	1989
Cultivated land ('000 ha)	196.6	199.6	214.0	205.3	231.2	212.7
(Paddy)	(46.4)	(56.8)	(63.2)	(122.5)	(129.1)	(136.0)
(Upland)	(150.2)	(142.8)	(150.8)	(82.8)	(102.1)	(76.7)
Grain production ('000 metric tons)	2 575	4 102	5 482	3 191	6 333	7 160
(Rice)	(1 350)	—	(2 159)	(2 122)	(3 603)	(5 898)
(Maize)	(333)	—	(2 681)	(22)	(60)	(121)
(Others)[a]	(892)	—	(642)	(1 047)	(2 670)	(1 141)
Farm population ('000)[b]	6 668	5 354	8 445	14 416	16 078	7 347
(% of total pop.)	(69.3)	(41.9)	(37.5)	(71.5)	(53.4)	(17.3)

[a] Soya beans and potatoes for North Korea; barley, wheat, soya beans, and potatoes for South Korea.
[b] In agricultural and fishery sector.

Sources: For North Korea's data, *Chosen Chungang Nyungam, Rodong Sinmun* (Pyongyang), and National Unification Board (Seoul); for South Korea's data, Economic Planning Board, Ministry of Agriculture and Fishery, and Bank of Korea yearbooks.

6.2.1 Agriculture

Table 6.2 shows total arable land and grain production in both North and South Korea. In North Korea the total arable land, including orchards, vegetables, and grain farms, is currently about 2 122 000 ha, of which approximately 10% is reclaimed land. About 80% of the arable land is located in the western part of the country. Rice paddies cover approximately 627 000 ha. Rice production per hectare is estimated to have been about 3443 kg, corn 3800 kg, and other miscellaneous grains 1300 kg in 1989.

In contrast, South Korea had about 2 127 000 ha (1.20 ha per farm family) in 1989, which had decreased from 2 312 000 ha (0.89 ha per farm) in 1967. This decrease was attributed mostly to the continuous encroachment on arable land for the building of highways and industrial factories. Of total arable land in the South, about 1 360 000 ha (that is, 64.4%) is paddy land. Rice production per hectare was estimated at 4690 kg in 1989.

The composition of the farm population in North Korea has dropped at a relatively lower rate than in the South over the last 45 years. The reason lies in the fact that the pragmatic North Korean leadership had established co-operative farms by the early 1950s, aiming both to achieve self-sufficiency in food and to fuel growth in industry. North Korean workers have not been free to change jobs without official permission, while South Korean rural workers have migrated in large numbers from the relatively stagnant primary sector to the rapidly surging secondary and tertiary sectors as the economy grows and transforms along the lines of Colin Clark's 'stages of development'.[1]

In passing, it must be noted that North Korea has achieved some success in increasing grain production through the Chungsan-ri spirit and method of farming. This refers to what Kim Il-Sung propounded as 'on-the-spot guidance' for agricultural work methods during his fifteen-day visit to the Chungsan-ri farm (now in the Kangso district of the Nampo municipality) on 5–20 February 1960.[2]

6.2.2 Manufacturing

Like other communist countries, North Korea has put most of its scarce resources into the development of heavy industry right from the beginning. This fostering of heavy industry did bring some real results in the early stages of development. North Korea's industrial sector more than doubled between 1953 and 1960, jumping to 62.3% from a meagre 30.7% of total output, as shown in Table 6.1. However, the singleminded emphasis on heavy industry caused not only the stagnation of light and consumer industry, but also crippling effects on the heavy industry itself by the failure to introduce advanced technology from overseas. In short, the inwardly oriented industrialization strategy based on the *chuche* philosophy appears partially responsible for the resulting drop in the import of new technology and also of productivity. Another drag on growth was the increasing complexity of the economy, making it more difficult for central planners to manage efficiently the allocation of resources. Bottlenecks inevitably developed, causing the affected factories to produce below capacity during the late 1960s and late 1970s.[3]

As a result, the pace of industrial growth—both heavy and light industry—in North Korea began to lag behind South Korea in terms of quantity, quality, and variety of goods produced. But this does not necessarily imply that all of North Korea's industrial sectors are being outstripped by those of the South. The North still has considerable potential in the areas of fertilizer and cement production, non-metallic mineral products, zinc ingot, lead products (pig iron and rolled steel),[4] and some machinery and equipment. South Korea is currently far ahead in most light industry and consumer goods, petroleum refineries, chemical products, electric and electronic equipment, iron and steel products, and car and shipbuilding industries.

Table 6.3 shows the production capability of North and South Korea's principal manufactures, although one-to-one mapping is not possible because of the different classification of products in the two countries.

Conscious of the consumer goods shortage, North Korea is increasingly stressing the necessity of boosting the supply of consumer goods by fully implementing the Workers' Party's policy concerning the light industrial revolution.[5] This policy refers to the 'on-the-spot' guidance that Kim Chong-Il, son of Kim Il-Sung, gave at the Light Industries Commodities Exhibition on 3 August 1984. Since then, production has expanded from some 4000 to some 10 000 kinds of consumer goods such as clothes, shoes, household items, and electronic goods.[6]

However, these consumer goods are mostly still produced by utilizing the North's own reserves from the scraps and by-products of heavy industrial production,[7] a reflection of the real shortage of resources for light and consumer industry development in North Korea.

6.2.3   Mineral Resources

North Korea is much richer than South Korea in deposits of mineral resources, with a few exceptions such as manganese ore, kaolin, silica stone, and silica sand. As shown in Table 6.4, the North overwhelms the South in known deposits of iron, lead ore, magnesium ore, bituminous coal, anthracite coal, tungsten ore, etc.

Given the abundance of mineral resources in the North and the heavy dependency of the South on imported resources, the two

TABLE 6.3 North and South Korea's principal manufactures, 1988 and 1989

	North Korea		South Korea[a]	
	1988	1989	1988	1989
Metal products				
Pig iron ('000 tons)	5 710	5 170	12 580	14 950
Steel, rolled ('000 tons)	5 040	5 940	3 540[b]	4 040[b]
Pressed steel products				
('000 tons)	3 970	4 040	5 040[c]	5 120[c]
Copper (tons)	90 400	90 400	234 385[d]	208 222[d]
Lead (tons)	87 500	87 500	21 674	19 397
Zinc (tons)	295 000	295 000	225 987[e]	241 743[e]
Aluminium (tons)	20 000	20 000	85 063[f]	77 331[f]
Machinery				
Passenger cars ('000)	18	33	868	846
Railway carriages	3 800	3 800	—	—
Engineering Machine ('000)	30	35	10	14
TV sets ('000)	240	240	14 820	15 180
Ships ('000 gross tons)	214	214	827	1 243
Construction and Chemicals				
Cement ('000 tons)	9 775	11 775	30	30
Magnetic cranks ('000 tons)	2 600	2 600	—	—
Fertilizer ('000 tons)	3 514	3 514	3 080	2 870
Fibres ('000 tons)	127	177	785[g]	844[g]
Cloth (million metres)	660	680	320[h]	290[h]

[a] Classification of products in South Korea differs from that of North Korea, making it difficult to map one-to-one in quantities.
[b] Concrete reinforcing steel bars only.
[c] Medium heavy plate and hot rolled sheet only.
[d] Copper wire bars only.
[e] Unwrought zinc ingots only.
[f] Aluminium plate only.
[g] Polyester and acrylic fibres only.
[h] Synthetic fibre fabrics only.

Sources: National Unification Board (Seoul), Bank of Korea (Seoul).

Koreas would greatly gain if they could complement one another in economic development. South Korea has an advantage in capital and technology, while North Korea has a plentiful supply of un-used resources. If inter-Korean trade could take place, both parties would be better off, at least by reducing transportation costs, unless any political transaction costs are taken into consideration.

TABLE 6.4 Estimated major mineral resource deposits in North and South Korea, 1989 ('000 tons)

	Estimated Deposits			Ratio (%)	
	North Korea	South Korea	Total	North Korea	South Korea
Uranium (U$_3$O$_8$)	26 000	56 000	82 000	31.7	68.3
Molybdenite (MoS$_2$)	2.0	32.5	34.5	5.8	94.2
Manganese (Mn) ore	200	1 250	1 450	13.8	86.2
Fluorite (CaF$_2$)	200	2 450	2 650	7.5	92.5
Pyrophyllite	125	10 000	10 125	1.1	98.9
Kaolin (Ka)	2 000	40 000	42 000	4.8	95.2
Silica sand (SiO$_2$)	6 600	31 613	38 213	17.3	82.7
Iron (Fe) ore	3 000 000	200 000	3 200 000	93.8	6.2
Lead (Pb) ore	12 029	640	12 669	94.9	5.1
Zinc (Zn)	12 000	0	12 000	100.0	0.0
Limestone (CaCo$_3$)	100 000 000	1 490 000	101 490 000	98.5	1.5
Magnesite (MgCo$_3$)	6 500 000	—	6 500 000	100.0	0
Anthracite coal (C)	11 740 000	1 450 000	13 190 000	89.0	11.0
Bituminous coal (C)	3 000 000	5 000	3 005 000	99.8	0.2
Nickel (Ni)	1 200	217	1 417	84.7	15.3
Gold (Au)	1	0.5	1.5	66.7	33.3
Tungsten (Wo$_3$)	232	185	417	55.7	44.3
Copper (Cu)	75	80	155	48.4	51.6
Graphite (G)	2 000	1 600	3 600	55.6	44.4
Talc	600	600	1 200	50.0	50.0
Beryllium (Be)	0.6	0.5	1.1	54.5	45.5

Sources: Korea Institute for Economics and Technology (Seoul), 'Mid- and Long-Term Development Plans for Economic Co-operation between South Korea and North Korea' (Draft), (April 1989), 73.

6.2.4 Foreign Trade Structure by Commodity Groups

As explained in Chapter 5, North Korea's foreign dependency ratio is less than 10% compared with South Korea's 60% in 1989. The volume of South Korea's foreign trade (exports plus imports) is about 25.8 times greater than that of North Korea. About 70% of North Korea's trade is with communist countries, of which about 55% is with the former Soviet Union and the remaining 15% with China. South Korea's trade, on the other hand, is heavily biased

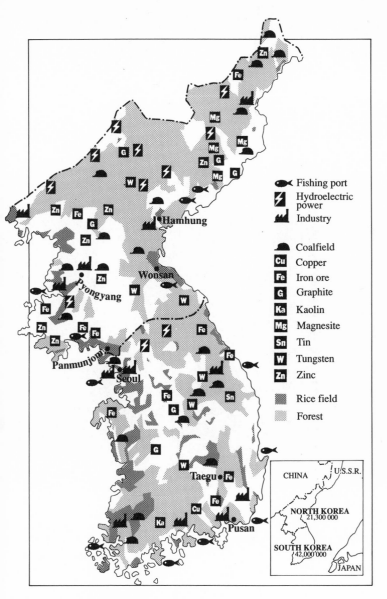

Resources of North and South Korea
Reproduced by courtesy of *Fortune* magazine.

TABLE 6.5 North and South Korean trade structures, by major commodity groups, 1987 (US$ million)[a]

SITC	North Korea[b]		South Korea[c]	
	Exports (f.o.b.)	Imports (c.i.f.)	Exports (f.o.b.)	Imports (c.i.f.)
0. Food and live animals	154.0	91.8	2 088.6	1 621.9
	(13.4)	(5.4)	(4.4)	(4.0)
1. Beverages and tobacco	6.4	5.0	89.7	32.1
	(0.6)	(0.3)	(0.2)	(0.1)
2. Crude materials, inedible	201.6	140.7	452.0	5 896.5
(excl. animal fodder)	(17.5)	(8.3)	(1.0)	(14.4)
3. Mineral fuels, lubricants,	72.6	581.1	747.5	6 021.7
and related materials	(6.3)	(34.4)	(1.6)	(14.7)
4. Animal and vegetable	0	8.7	4.4	138.9
oils, fats, and waxes	(0.0)	(0.5)	(0.0)	(0.3)
5. Chemicals and related products	20.8	76.7	1 320.7	4 594.7
	(1.8)	(4.5)	(2.8)	(11.2)

6. Manufactured goods	395.4	263.1	10 197.8	6 255.0
	(34.4)	(15.6)	(21.6)	(15.2)
7. Machinery transport equipment	119.7	460.0	16 906.0	13 911.3
	(10.4)	(27.2)	(35.8)	(33.9)
8. Miscellaneous manufactured articles	175.8	47.3	15 348.8	2 139.6
	(15.3)	(2.8)	(32.5)	(5.2)
9. Other commodities	4.0	14.4	125.4	408.1
	(0.3)	(0.9)	(0.0)	(1.0)
Total	1150.4	1688.7	47 280.9	41 019.8
	(100.0)	(100.0)	(100.0)	(100.0)

[a] Based on UN, SITC (revised); figures in parentheses are percentages.
[b] North Korea's trade data are calculated as follows:
(i) North Korean trade with Soviet Union, China, OECD, Hong Kong, Singapore only (other countries excluded). Therefore, figures here for 1987 are not exactly the same as those in Table 5.1 above.
(ii) Exchange rates used are: US$ 1 = 0.627 OECD currency = 3.7221 Chinese *won* = 7.798 Hong Kong dollars = 2.106 Singapore dollars.
(iii) c.i.f. data are divided by 1.1, and f.o.b. data are multiplied by 1.1, to derive North Korea's imports and exports.

Sources: Bank of Korea, *Economic Statistics Yearbook* (1990); OECD Trade Magnetic Tape; Customs General Administration of China, *Summary Surveys of China's Customs Statistics* (1988); *Hong Kong Trade Statistics* (1988); *Singapore Trade Statistics* (1988); USSR statistics; see also KIET source for Table 6.4, pp. 94–5.

towards the United States (30%) and Japan (25%). The Northern Triangle relationship (Russia–DPRK–China) exists parallel to the Southern Triangle relationship (USA–ROK–Japan), even though the absolute sizes of these relations are greatly different in terms of trade volumes and values.

North Korea exports mostly primary commodities, mineral resources, and some metallurgical, chemical, and engineering products, and imports manufactured goods, while South Korea exports manufactured goods and largely imports raw materials and intermediate goods. Both countries export and import machinery, transportation equipment and other manufactured goods. This illustrates that, were inter-Korean trade to be established, the two Koreas have the potential to establish a vertical co-operative relation in raw materials trade as well as a horizontal relation in the exchange of different manufactured goods, to the mutual comparative advantage of both.

Table 6.5 compares the trade structures of North and South Korea by major commodity groups.

South Korea's 'northern policy' has recently succeeded somewhat in opening up diplomatic and trade relations with most communist bloc countries. For this reason, relations between North Korea and Western countries are likely to take a new turn sooner or later. Such a development will provide further momentum for the two Koreas to seek to establish a dialogue and other positive interaction including inter-Korean trade and cultural exchanges.

Indeed, a first step of inter-Korean trade through indirect channels has already started, following South Korea's proposals, made unilaterally in October 1988, for an exchange of commodities. At the end of March 1992, the total value of two-way trade between North and South Korea via third countries was estimated at $168 100 000 on an actual clearance basis. It is known that South Korea buys steel products, lead, zinc ore, agricultural and fishery products, and Chinese medicine herbs from the North. In return, the South sells jackets, sportswear, cigarette filters, and sugar to the North. Table 6.6(*a*) shows South Korea's permit issues for the private exchange of commodities with North Korea via third countries from 1988 to the first quarter of 1992. Tables 6.6(*b*) and 6.6(*c*) record the total value of actual transactions between North and South on a customs clearance basis and South Korea's total trade with North Korea by commodity groups at March 1992.

TABLE 6.6(*a*) South Korea's inter-trade permits, firms, and value, 1988–1992

	No. of permits	No. of firms	No. of commodities	Value (US$ '000)
Imports				
1988	4	4	16	1 037
1989	57	47	53	22 235
1990	75	60	88	20 354
1991	328	233	200	165 996
1992 (I)	117	81	85	59 104
	581	425	442	268 726
Exports				
1989	1	1	1	69
1990	4	4	4	4 731
1991	40	16	58	26 176
1992 (I)	17	9	8	6 367
	62	30	71	37 343
Total	643	455	513	306 069

Sources: Ministry of Commerce and Industry (Seoul); National Unification Board (Seoul), 1992 (IV).

Before we consider some further approaches to direct trade and exchange between North and South Korea, it seems appropriate to review briefly the history of contacts between the two states, followed by a review of the wisdom and experience of economic interaction between the different economic systems of the former East and West Germanys.

6.3   THE OPENING OF NORTH–SOUTH DIALOGUE

A first attempt at one-to-one talks between the two rival Koreas began in 1971 after more than two decades of frozen hostility. Park Chung-Hee, President of the Republic of Korea, issued a challenge to North Korea in a speech on 15 August 1970 proposing to remove man-made barriers between North and South.

TABLE 6.6(*b*) Actual transactions between North and South Korea (customs clearance basis), 1989–1992

	No. of transactions	No. of firms	No. of items	Value (US$ '000)
Imports				
1989	66	55	55	18 655
1990	78	60	60	12 278
1991	300	184	153	105 722
1992 (I)–(III)	104	55	51	21 258
	548	354	319	157 913
Exports				
1989	1	1	1	69
1990	4	3	6	1 187
1991	23	16	21	5 547
1992 (I)–(III)	14	9	6	3 384
	42	29	34	10 187
Total	590	383	353	168 100

Sources: as for Table 6.6(*a*).

At the North Korean Workers' Party's Fifth Congress in November 1970, President Kim Il-Sung brushed aside Park's proposal as tactical and political propaganda and emphasized the need to overthrow the fascist military dictatorship in the South and establish a true people's government.[8] In April 1971, North Korean Foreign Minister Ho Tam came up with a comprehensive eight-point programme for peaceful unification.[9]

In the midst of the subsequent exchanges of proposals and counter-proposals, conducted in the harsh rhetoric used by each side towards the other, 25 preliminary Red Cross talks had progressed, between 20 September 1971 and 11 August 1972, to an agreement on arrangements for the first full-dress meeting of the Red Cross Conference to begin in Pyongyang on 29 August 1972. Behind the scenes of the preliminary Red Cross talks, private contacts between the two leaders resulted in secret visits in May–June 1972 to Pyongyang by Lee Hu-Rak, director of the Korean Central Intelligence Agency, and to Seoul by Deputy Prime

Minister Park Song-Chol. Lee Hu-Rak met Kim Il-Sung and Park Song-Chol met Park Chung-Hee. The result was a North–South joint statement on 4 July 1972, signed by Lee Hu-Rak and Kim Yong-Ju (Kim Il-Sung's younger brother and director of the Organization and Guidance Department of the KWP Central Committee) on behalf of their 'respective superiors', expressing agreement on three 'principles for unification of the fatherland'.[10]

Since then, however, no substantial progress had been achieved between the two sides, with the former provocations merely being replaced by 'new confrontation with a dialogue'. After Park Chung-Hee had announced the ROK's new foreign policy on 23 June 1973, expressing the opinion that temporary membership of both Koreas in the UN would be necessary for further dialogue, Kim Il-Sung angrily broke off the formal talks on 28 August 1973. North Korea even refused to continue the Red Cross working-level talks in March 1978, charging that the Team Spirit military exercise made such meetings impossible.

On 23 June 1978, Park Chung-Hee again proposed, on the occasion of the five-year anniversary of the first proposal, the formation of a consultative body for promoting North–South economic co-operation, and offered to hold a ministerial conference with North Korea. The two sides then proceeded to exchange a series of counter-proposals and preconditions with no discernible progress until 1984.

On 20 August 1984, President Chun Doo-Hwan proposed inter-trade and economic co-operation to North Korea, explaining that the South was willing to provide the North with materials and technology to raise the people's living standards. Shortly after that, Seoul and other areas in the South were flooded with record-high rainfalls, leaving hundreds of victims of the calamity without food and shelter. On 14 September 1984, North Korea's Red Cross promptly offered South Korea's flood sufferers aid, and the offer was reluctantly accepted by the South. Rice and other relief goods arrived via Panmunjom (a truce village) and the ports of Inchon (west coast) and Pukpyong (east coast) between 29 September and 4 October 1984. Since then, talks about economic co-operation and trade, Red Cross talks about family reunions and cultural and sports exchanges, and higher-level talks have been off and on, still fostering hopes of paving the way for Korean *râpprochement*.

On the basis of President Roh Tae-Woo's special declaration

TABLE 6.6(c) South Korea's trade with North Korea by major commodity groups, 1989–1991 (US$ '000)

1989		1990		1991	
Item	Value	Item	Value	Item	Value
Imports					
Zinc ore	5 701	Potatoes	4 195	Zinc ore	47 047
Thermocurrent coil	5 700	Zinc ore	3 026	Gold ore	17 537
Electrolytic copper	2 447	Cement	1 541	Thermal coil	9 136
Raw silk	1 311	Other	3 516	Steel billets	7 608
Anthracite coal	1 050			Cement	3 394
Other	2 446			Anthracite coal	2 779
				Herbs	2 593
				Frozen pollack	1 718
				Raw silk	1 588
				Lead ore	1 466
				Silver ore	1 204
				Other	9 652
Total	18 655		12 278		105 722

Exports

Clothes (knitwear)	69	Socks	1 094	Rice	1 607
		Cigarette filters	83	Diesel oil	1 392
		Sugar	10	Plasticizer	1 137
				Colour TVs	427
				Polyethylene	250
				LDPE	216
				PS resin	161
				HDPE	146
				Catalyzer	56
				Cigarette papers	40
				Refrigerators	20
				Polyester textiles	18
				Polypropylene resin	10
				Other	23
Total	69		1 187		5 547

Sources: as for Table 6.6(a).

for national self-esteem, unification, and prosperity on 7 July 1988, the South Korean government announced on 7 October 1988 its limited permission for private domestic businesses to trade commodities and raw materials with North Korea, and also permitted North Korean commercial ships entry into ports in the South. At the same time, South Koreans were allowed to visit most communist countries, and domestic firms were encouraged to promote mutual trade with those countries.[11]

On 1 January 1990, President Kim Il-Sung proposed that North and South Koreans be allowed free travel between the two parts of Korea. He said, in his 1990 New Year message:

The North and the South should not confine their efforts to guaranteeing free travel, but should proceed to opening their doors fully in all spheres, including politics, the economy, and culture. In order to remove the barrier of national division and realize free travel between the North and the South and an open-door policy, negotiations must be held without delay. For this purpose, we propose a top-level North–South conference in which the heads of the authorities and the leaders of political parties will take part.[12]

On 20 July 1990 President Roh Tae-Woo of South Korea announced that 'during the five-day span around 15 August, the 45th anniversary of national liberation, the entire region of South Korea will be open to all North Korean residents for free visits, and North Korean authorities are urged to take reciprocal measures'.[13]

The proposal was not taken up, with earlier confrontations between Seoul and Pyongyang involving a veritable tug-of-war over the issue of whether Pyongyang's delegates would come to Seoul to attend the 'second preliminary meeting' to arrange for a Pyongyang-sponsored pan-national rally to be held in the truce village of Panmunjom on 13–17 August. However, such proposals did lead to talks between the two Korean presidents, held first on 4–6 September 1990 in Seoul, and then on 16–19 October in Pyongyang, and again in Seoul in December 1990.

Thus, the recent situation on the Korean peninsula would seem to be associated with unprecedented vigour on the part of both Seoul and Pyongyang in promoting various inter-Korean exchanges. In addition to the first-ever high-level talks held alternately in Seoul and Pyongyang, inter-Korean soccer games were held in Pyongyang (11 October 1990) and Seoul (23 October 1990). North Korea also hosted the first pan-Korean Music Festival for Reunification in Pyongyang on 18–23 October 1990. Finally, both presidents signed a four-chapter basic agreement in Seoul on 13 December 1991, entitled an 'Agreement Concerning Reconciliation, Non-aggression, Exchange and Cooperation Between the North and the South'. Although any discernible progress is hard to expect from such ongoing North–South dialogue and other contacts in the near future, because of the differences of intention between the two sides, the continuation of meetings between the two ideologically opposed countries will be rewarding in the long run. Moreover, the German model of inter-country trade and of other East–West trade relations may shed some insights on future programmes of inter-Korean ties.[14]

6.4 GENERAL LESSONS FROM THE EAST–WEST GERMANY INTERACTION

In that time he learned that the world is all one piece. He learned that the world is like an enormous spider web and if you touch it, however lightly, at any point, the vibration ripples to the remotest perimeter and the drowsy spider feels the tingle and is drowsy no more but springs out to fling the gossamer coils about you who have touched the web and then inject the black, numbing poison under your hide. It does not matter whether or not you meant to brush the web of things.

'All the King's Men' by Robert Penn Warren[15]

6.4.1 General Problems

The foreign trade of any country is closely related to the political and economic structure of that country. The trade pattern of a centrally planned economy (CPE) differs significantly from that of a market-oriented economy, both in the type of goods (and services) traded and in how such trade is carried out.

The distinguishing feature of communist economies like North Korea is the virtual absence of private enterprises which are the core players in the trade of market economies. Furthermore, communist economies have an hierarchical organization in which decisions emanate from a central authority. Consequently, there is a large degree of centralization of decision-making.[16]

The decision-makers in centrally planned economies such as North Korea determine at their discretion what to produce and how to produce it, and the prices that are set play little or no role in output and material supplies. Profit-oriented price competition seldom exists in such a planning system. It is not the prices on output but the decision determining what material supplies can be obtained under the plan that is most important. In market economies, on the other hand, the price mechanism is the most important factor determining supply and demand, although some sort of hybrid indicative planning, involving the regulation of prices, wages, taxes, and interest rates, is very often practised.[17] It must be remembered that under indicative planning, producers determine what to produce, in what amounts, and to whom they will sell, but they do not have the right independently to set basic prices on the goods they produce or the resources they use. Such standard indicative planning is what is mostly used by mixed-economy socialist countries, while capitalist countries impose restrictions only on the prices of some selective goods and resources, often public goods and those produced by 'natural' monopolies.

The planners in the CPEs often use 'incentive methods' to encourage hard work and increased production. However, in a system of centrally allocated supplies and planned production levels not all economic incentives will produce results, because money in these economies is simply an accounting unit. Indeed, money can be a real incentive only when it can be used to purchase goods in a situation of no quantity constraints. In a planned economy such as that of North Korea, where economic goals are set entirely by the central government, the incentive system usually ends up affecting the extent to which managers and workers follow their superiors' orders. Because it is the quantity of output, rather than consumer satisfaction, that is emphasized, there will be a lack of focus on quality, product variety, and other less easily quantifiable facets of economic activity. Managers in CPEs such as North Korea are more isolated from customers and suppliers than firms in market economies. Often they liaise only with their superiors, avoiding contact with those of their own grade of responsibility. This remoteness from agents on the same level of the hierarchy leads to poor information flows, especially with regard to the specific needs of customers and the availability of technological opportunities.

The remoteness from customers and suppliers has been especially important in international activities.[18] The Ministry of Foreign Trade is akin to any other ministry in a communist country. It aids in the planning process, channels information about the activities of its operational units of the planners, implements the relevant part of the plan by disaggregating the plan figures, and supervises a group of producers.[19] However, the Soviet-type planned systems usually lack an *ex ante* link between the producers and users of internationally traded goods on the one hand and the enterprises that have the job of buying and selling on foreign markets on the other. The decision-making overload at the centre, with the consequent need to reduce the amount of information-processing, probably accounts for lack of *ex ante* co-ordination between foreign trade enterprises and producing enterprises. Fig. 6.1 shows the working organization of foreign trade planning in a typical CPE.

One can draw a few general lessons concerning the structure of trade from a knowledge of the typical internal organization of foreign trade in the CPEs. The presence of a planning hierarchy and the absence of profit-oriented private firms are the most distinguishing features. Any comparative analysis of these characteristics with regard to international trade is beyond the scope of this study. However, the vastly different features of the opposing economic systems certainly go to make interaction between two such economies difficult. Before initiating any full-scale reforms in terms of price systems and industrial organization in either or both of the systems, there is the question of how to promulgate inter-country trade between North and South Korea by minimizing conflicts. To gain some insights into that problem, we shall examine trade between the two Germanys as well as their intra-industry trade experiences, which it is thought have aided the German reunification.

6.4.2 West Germany's Trade and Economic Co-operation with the GDR

The political earthquake that wrought the 1989–90 collapse of East Germany and started the move towards reunification with West Germany may trace its seismic centre back to former West German Chancellor Willy Brandt's *Ostpolitik* (Eastern policy), a policy of small steps taken to accommodate the German Democratic Republic (GDR). Ever since Brandt initiated *Ostpolitik* in

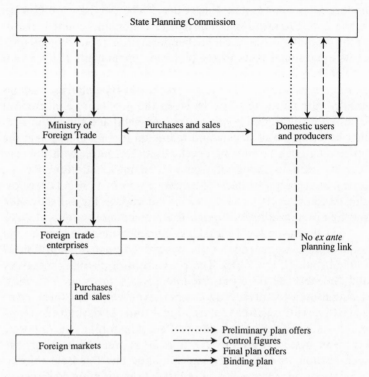

FIG. 6.1 Organization of foreign trade in a Soviet-type economy
Source: Frank D. Weiss, *West Germany's Trade with the East: Hypothesis and Perspectives,* Kieler Studien no. 179 (J. C. B. Mohr, Tübingen, 1983), 24.

1969, successive West German governments quietly but increasingly adopted internal and external policies aimed at ameliorating the costs of partition. The small steps of *Ostpolitik* gradually gave way to more headlong strides toward reunification, through contacts, reconciliation, and an acceptance of what then seemed to be the reality of communist rule.[20]

Following Willy Brandt's first visit to the GDR to hold a high-level talk with East Germany's Erfurt on 19 March 1970, the relationship between the two Germanys began to improve. As a consequence, a basic treaty for the normalization of relations between East and West Germany was concluded on 21 December 1972. The two Germanys concurrently became members of the

United Nations in September 1973, and the two governments established diplomatic offices in Bonn and East Berlin in March 1974.

Willy Brandt's *Ostpolitik*, Helmut Schmidt's attempts to preserve *détente* in the early 1980s in the wake of crises in Afghanistan and Poland, and the emotional debate over intermediate-range nuclear forces, the double-zero option, and concerns that Germany would become a singularly powerful state created a new awareness of unique German interests, interests increasingly seen as transcending the East–West divide.[21]

Inter-German trade and economic co-operation had already begun on an individual basis as early as 1946, long before the official normalization in 1972. Intra-industry trade occurred when West German citizens were allowed freely to visit their relatives in the East. Hard currency flooded into East Germany both through remittances from West German relatives and through other means, including financial support from West Germany.

The chronology of mutual co-operation between the two Germanys can be classified into three stages as follows:

First stage	1951–89	Inter-German trade
Second stage	1970	Communication and transportation
	1972	Exchange
	1975	Common measure for environment issues
	1976	Joint mining exploitation
	1979	Cultural exchange
	1982	Exchange of private organizations
	1983	Mutual loans and grants offer
Third stage	1987	Science and technology co-operation
		Environmental protection co-operation
		Cultural exchange co-operation
	1989	Opening of the Berlin Wall and inter-German borders
	1990	German reunification

Since 1960, visible trade between the two Germanys had grown at an annual average rate of 9%. As shown in Table 6.7, total two-way trade between East Germany and West Germany grew from DM745 million in 1950 to DM15 306 million in 1989. The

TABLE 6.7 Trade between East and West Germany, selected years, 1950–1989

	Two-way total trade (DM million)	% of total West German trade volume	% of total East German trade volume
1950	745	3.8	20.3
1960	2 082	2.3	11.3
1970	4 411	1.9	11.1
1980	10 873	1.6	9.0
1989	15 306	1.3	20.6

Sources: Statistisches Bundesamt, *Statisches Jahrbuch* (1984, 1990).

inter-German trade proportion of West Germany's total external trade dropped from 3.8% in 1950 to 1.3% in 1989, while the share of inter-trade of East Germany's external trade accounted for 20.6% in 1989. Indeed, West Germany had remained a very important trade partner of East Germany, second only to the Soviet Union, ever since 1950.

West Germany exported crude oil, steel products, chemicals, machinery, and electric and electronic products to East Germany, while importing petrol, diesel fuels, heating oil, chemicals, steels, and textile products during the 1970s and 1980s.

Besides visible exports and imports, the two Germanys managed to maintain economic co-operation throughout the post-war period, most of which was in terms of flows of economic assistance from West to East. They included a DM2000 million subsidy for construction and repair of roads linking West Germany and Berlin, DM2000–3000 million of payments to help cover expenses for resettlement of political offenders in East Germany, and DM1950 million of loan guarantees for East Germany. About DM4000 million worth of goods or cash had flowed annually from West to East in various types of economic co-operation. This aid obviously helped East Germany survive as the most efficient and durable communist regime in Eastern Europe until the revolution in autumn 1989. Table 6.8 records West German assistance to East Germany during the period 1975–88.

TABLE 6.8 West Germany's economic assistance to East Germany, 1975–1988 (DM billion/VE[a])

Items	Transaction in settlement accounts	Transaction in international currency	Total value
I Current balance	+2.7	−25.1	−22.4
(A) Goods and services	+2.8	−10.2	−7.4
Goods	−4.4	+4.6	+0.2
Services	+7.2	−14.8	−7.6
(B) Transfer payments	−0.1	−14.9	−15.0
(a) Private sector	−0.1	−2.9	−3.0
(b) Public sector	—	−12.0	−12.0
Toll and travel fees	—	−6.6	−6.6
Subsidy for East Germans visiting West Germany	—	−2.0	−2.0
Others	—	−3.4	−3.4
II Capital balance	−2.2	+0.2	−2.0
III Current and capital balance	+0.5	−24.9	−24.4
IV Unclassified			−35.5
V Total funds transferred to East Germany (III + IV)			−59.9

[a] 1 DM (West German mark) = 1 VE (Verrechnungseinheit) = 1 mark (East German). A minus sign indicates negative balance on the part of West Germany's account.

Source: *Monthly Report* of Deutsche Bundesbank (January 1990).

With respect to the settlement methods of *Innerdeutscher Handel* (inter-German trade), the two Germanys agreed in 1972 to use the *Verrechnungseinheit* (VE); this was established on the fixed exchange rate of DM 1 = VE 1 = EM 1 (GDR *ostmark*). Payments for *Innerdeutscher Handel* were made annually through mutual settlement accounts (*Konto*) set up in the respective central bank of each country. To complement the settlement account methods, however, the two Germanys established a credit system called *swing* (overdraft), which enabled the trade deficit side to draw credit within mutually agreed-upon limits at no interest. When the trade deficit exceeded the annual *swing* limits, either a supplier's credit or a bank loan had to be arranged. The limits of annual *swing* overdraft were adjusted by mutual agreements.

Along with enhanced inter-country trade and economic co-operation between the two Germanys, the 1972 general agreement on free travel across the inter-German border accelerated the number of mutual visitors from West to East Germany. The number of West Germans visiting the East increased from 2 600 000 in 1970 to 6 200 000 in 1972 and to 7 700 000 in 1975.

The expansion of material and human exchanges served to help the two German states grow together again in a broader sense. Interaction between the two political systems ensured that Germans in the GDR could discover the virtues of the Western-style freedom and prosperity that their West German brethren had long enjoyed. From within, hidden forces emerging out of the mutual interaction and exchange were seen as working increasingly towards the disintegration of the rigidly controlled East German government. Of course, the primary motive behind such developments in the GDR was the desire for Western standards of political freedom and economic affluence. Such demands could scarcely have been met in East Germany had the rising discontent there not been accompanied by the Gorbachev era of *perestroika*, which marked a crucial shift in the entire political terrain of Eastern Europe by autumn 1989.

After the toppling of the Honecker regime in October 1989, followed by the crumbling of the Berlin Wall in early November, German reunification appeared to be inevitable. On 7 February 1990, West German Chancellor Kohl offered a monetary union with the introduction of the Deutschmark in East Germany, conditional on simultaneous economic reform. In spite of many problems of inequality in the political and economic systems, infrastructures, regional productivities, living standards, etc., which are being encountered in the transition to one economy, the two Germanys dropped the 'East' and 'West' from their names to become once again a single nation on midnight of 1 October 1990, with the ringing of the Liberty Bell at Schöneberg Town Hall in Berlin.

6.4.3 The Unequal Brothers: Germany and Korea

There are similarities as well as differences between the two Germanys and the two Koreas. The most important similarity is that the market-oriented and mixed economies[22] overwhelm the centrally planned economies in both economic growth and trade achievement. The most visible cross-country dissimilarity can be found in the fact that the two Germanys had kept up various contacts between one another for a long time before unification, while the two Koreas have not been in touch with one another very much until recently. One can also find some disparity in the differing degree of economic strengths and political stability in West Germany and South Korea; on the other hand, the two orthodox communist regimes in North Korea and East Germany had much more in common, both politically and economically.

Another important point of comparison is that the structure of the East German industrial sector is very similar to that of West Germany, while that of North Korea differs in so many respects from that of South Korea. East Germany was dominated by mechanical engineering, chemicals, electrical engineering, precision tools, and optics (see Fig. 6.2). One of the largest branches of industry in East Germany—as in West Germany and also North Korea—is mechanical engineering. Most of East Germany's output—again, like North Korea—was exported to the Soviet Union. In 1988, East Germany's trade with the Soviet Union ranked in first place and accounted for about 40% of total East German trade.

With reunification, Germany's population is almost 80 million, or about a quarter more than the 1989 population of West Germany alone. But with its low standard of living, East Germany adds only 10% to West Germany's gross national product.[23]

Table 6.9 compares recent economic indicators between East and West Germany and between North and South Korea.

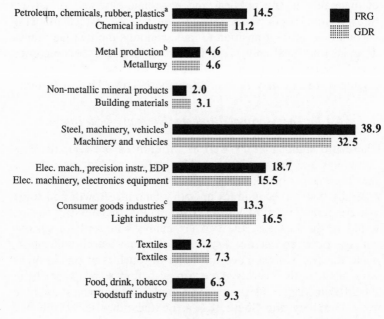

^aIncluding mining.
^bIncluding foundries, light metal products.
^cIncluding wood processing, excluding textiles.

FIG. 6.2 West and East German employment in industry, 1988 (%)
Source: Federal Statistical Office, *Statistical Yearbook of the GDR* (1988); recopied from Wili Leibfritz, 'Economic Consequences of German Unification', *Business Economics*, 25:4 (1990), 7.

While per capita incomes alone do not explain the differences in the quality of production or in the real welfare level between the two systems, perhaps the most striking comparison is the industrial performance of East and West Germany. At the end of the Second World War, the two Germanys had roughly equal levels of productivity and similar industrial structures. After four decades of capitalism in the West and communism in the East, per capita income or productivity in East Germany had fallen to a level estimated at between one-third and two-thirds that of West Germany, depending on the exchange rate used. In contrast, overall productivity (in terms of per capita income) in North Korea had

| | Germany | | Korea | |
	East	West	North	South
Population, 1989	16 400 000	62 079 000	22 521 000	42 793 000
GNP, 1989 (US$ million)	44 000~95 000[b]	1 200 000	25 800~55 900[d]	210 100
Per capita GNP, 1989 (US$)	2690~5810	19 330	1146~2481	4968
Industrial productivity, 1990	53	100	30.8[e]	100
Foreign trade, 1989 (US$ million)				
Exports	13 600	341 110	1560	62 380
Imports	14 300	269 500	2520	61 470
No. of people per telephone (per car), 1986–8	4.3 (4.8)	1.6 (2.2)	8.9 (80.0)	5.4 (50.7)
Average monthly wage, 1989 (DM for Germanys, US$ for Koreas)	1016	3469	75~150[d]	700
GDP from agriculture, 1988 (%)	12.9[f]	1.5	24.0[g]	10.8[h]
Labour force in agriculture, 1988 (%)	8.6	3.9	35.3	26.8
Population per sq. km. of arable land, 1988	354	1303	952	2124
Human development index (1990)[i]	95.3	96.7	78.9	90.3

[a] US$1 = DM8.14.
[b] US$1 = DM3.76.
[c] US$1 = W2.08.
[d] US$1 = W0.96.
[e] Electricity generation ratio with South Korea = 100.
[f] As a % of net material product.
[g] 1981.
[h] 1987.
[i] This index is constructed by the UN Development Programme (UNDP) by using life expectancy, literacy, and sufficiency of average income estimates meeting basic needs. Countries scoring less than 50 are classified as having low human development, those scoring 50–80 as medium, and those above 80 as high.

Sources: Federal Statistical Office (West Germany), *Statisches Jahrbuch* (1990); Bank of Korea, *Statistical Yearbook* (1990); UNDP (1990); 'Vital World Statistics', *The Economist* (1990), 56 and 228 for GDP and labour force in agriculture, and human development index, and 234 for number of people per telephone (and per car).

fallen to a level of one-half of that in South Korea by 1979, if the trade exchange rate is used as a conversion factor (see Chapter 3). However, it is advisable to note that this rate will tend to under-estimate real incomes because of the omission of non-tradable goods.

Cross-country differences between the two Germanys and the two Koreas can also be found in the following facts. First, the Germans took a much more practical and humanistic approach to-wards establishing relations. They started trading with one another on a private basis as early as 1946, one year after the division. In that sense, Koreans are more than four decades behind the Germans in terms of inter-country trade and exchanges. Second, the two Koreas have maintained an active hostility towards one another ever since the Korean War, and they have followed completely different political ideologies and socio-economic structures. The gap between North and South is a result of the complete break in relations for more than 40 years, and will not easily be bridged. Third, the immediate target of West Germany's *Ostpolitik* was East Germany, while South Korea's recently ini-tiated 'northern policy' has as multiple targets Russia, China, Mongolia, *and* North Korea. Considering the stubbornness and rigidity of the North Korean regime, this circuitous approach is perhaps ill advised at this time.[24] Meanwhile, both sides are cur-rently stepping up political propaganda campaigns against each other amid the expanding inter-Korean exchanges, raising serious doubts about the political leaders' real intentions.[25] Therefore, the two Koreas need to establish mutual trust and confidence above all before taking steps towards practical co-operation and inter-country trade, let alone reunification.

The overnight reunification of the two Germanys and the ongoing economic reforms of many former Eastern European communist countries are putting pressure on the North Korean leadership. South Koreans now need to take a more conciliatory approach towards the North. North Koreans, for their part, need to reconsider their attitude towards the South if they seriously intend to work towards North–South *râpprochement*.

Whether Koreans like it or not, the interaction and reunification of East and West Germany is the only such experience to date concerning two ideologically divided states. The reunited Germans are still struggling to overcome their contrasting experiences as citizens of two very different countries. There are many lessons for Koreans to learn from studying how the two German populations have had to compromise with one another and how they are responding to the challenges created by their economic and political reunification. Of course, all this can best be understood only in relation to Germany's own internal and external dynamics. Indeed, inter-German relations evolved over several decades of inter-country trade and co-operation, with a view to a homogeneous, harmonious, low-key, and nationalistic *râpprochement*. Their reunification is a remarkable success for all Germans, whatever the costs and problems involved (see Table 6.10), and Germany's continuing struggle thus continues to be a matter of international interest and one that will provide the divided Koreans with new thinking on the Korean Question.

6.5 THE FUTURE OF INTER-KOREAN RELATIONS AND UNIFICATION

6.5.1 A Preface

Until late 1990, it looked as if the DPRK had the only durable—and most hard-line—closed regime in the communist bloc. North Korea's newspapers and radio have severely criticized and scorned the reform policies of both the former Soviet Union and the Eastern European countries, pointing out that 'capitalist and imperialist ideology and culture paralyses the revolutionary consciousness in those countries'.[26]

North Koreans had maintained closer relations with the Soviets than anyone else in the communist bloc until the USSR entered into diplomatic ties with Seoul. In a broader sense, North Korea has much resembled the GDR in terms of the toughness and the orthodoxy of its Marxism–Stalinism. In the post-war era, however, the East Germans became more servile to their Russian masters than anyone in the bloc—as countless bitter jokes attested.[27] They also became the most prosperous member in the communist bloc. East Berliners scorned the Hungarian uprising of 1956, the Prague Spring of 1968, Solidarity's strike of 1980, and the Polish and Hungarian reforms of 1989 for their romantic sacrifice of an already poor standard of living for a political will-o'-the-wisp. East Germans congratulated themselves on their superior realism and

TABLE 6.10 Estimated cost of German unification, 1990–2000

		DM100 million
(I)	Social welfare costs	5010
	Unemployment compensation[a]	2560
	Medical insurance subsidy	700
	Retirement pensions[b]	250
	Budgetary aid to East Germany	1500
(II)	Economic rehabilitation expenses	8000~11 000
	Investment for roads and railways	2000
	Housing construction	2000
	Environment improvement	3000–5000
	Expansion and modernization of communication facilities	1000–2000
(III)	Total ((I) + (II))	13 010~16 010

[a] Assume 20% unemployment rate.
[b] Assume 50%~70% the level of West Germany.
Source: David C. Roche, *Der Spiegel*, no. 42 (1990).

cynicism.[28] In addition, the GDR's internal control system—like that of the DPRK—worked well, it seemed, and could perpetuate itself indefinitely. Even the advent of radical reformer Mikhail Gorbachev in the Soviet Union failed to alter East Germany's internal stasis, just as, now, it does not succeed in changing DPRK's leadership. The former GDR, which was relatively stable, was in many ways the least likely country in Eastern Europe to undergo radical political change; but it no longer exists in a World Atlas of 1993.

Today North Korea, one of the few survivors in an orthodox communist system, would seem to be the most unlikely country to dissociate itself from its 'superior communism'. Nor are South Koreans likely to accommodate the autarkic communist system. Given that there continue to be two ideologically opposed Koreas in the peninsula, the common task facing them seems to be to seek to diminish the existing antagonism between the two halves in an effort to attain *râpprochement*.

For the two Koreas, the transformation may not come in such a revolutionary fashion as it did in Germany. Perhaps, though, gradualism is the best approach to 'one Korea'. Leaving 'champagne and tears' for the distant future, the two Koreas should now work to find ways of overcoming their poor relations so that they can begin to grow together towards reunification. The task will never be easy, because more than four decades of division have sent them along very different roads, and because compromise between Stalinist communism and capitalism is never easy, just as it is not easy to mix oil and water.

However, where there is patience and willingness to compromise, there is still hope. The hope emerges when the two Koreas find that they are not fundamentally so different after all. Both sides will come to understand this 'sameness' and their basic heterogeneity if they can make further progress in the North–South dialogue, perhaps establishing exchanges of soccer teams, musicians, and high-level political groups.

First of all, both sides must try to continue the various levels of face-to-face meetings in all candour, surrendering all objectives of discrediting the other party. The importance of understanding and perseverance on both sides cannot be overemphasized. Inter-Korean high-level talks must proceed steadily so as to build up mutual confidence and to identify areas of agreement as well as differences in the bilateral positions. These talks may be accompanied or followed by inter-Korean economic co-operation and other exchanges to narrow the chasm during a period of gradual transition to national unity.

The remainder of this chapter will consider possible efforts at economic co-operation between North and South Korea.

6.5.2   Towards Inter-Korean Economic Co-operation

The need for establishing inter-Korean relations has been raised many times by North and South Korean leaders alike. The first call for a prime ministers' conference was forwarded by Pyongyang on 11 January 1980, when the South was in political uncertainty in the aftermath of the assassination of President Park Chung-Hee in October 1979. A similar proposal was made by the North, on 10 February 1982, for a joint meeting between North and South Korean politicians involving 50 prominent delegates from each side. None of these plans materialized, however, because of several domestic and external events, including the Rangoon bombing of 1983 and a KAL explosion just before the 24th Olympic Games in Seoul in 1988.[29]

With the rapidly changing international environment, President Roh Tae-Woo unveiled a new foreign policy initiative towards North Korea and other communist countries on 7 July 1988. Subsequently, Roh called for a summit meeting between himself and President Kim Il-Sung. President Kim, who was certainly aware of and sensitive to such new developments as *détente* between the East and West as well as South Korea's application for admission to the UN in autumn 1990, countered by proposing the prime ministers' talks again in his 1990 New Year address. This proposal was accommodated in due course by the South, and the historical four-day inter-Korean prime ministers' talks began on 4 September in Seoul, with a 90-member delegation from Pyongyang visiting via the truce village of Panmunjom, as described above in Section 6.3.

Expectations of economic co-operation between North and South Korea soared higher than ever, at least on the surface, in the wake of these prime-ministerial talks. The interest of both sides seemed to herald significant changes in inter-Korean relations, with North Korea seeming to seek to open its doors. The Northern leadership is apparently conscious of the threat to their power if this happens. They are also well aware of the dilemma that they must risk further economic stagnation if North Korea remains closed.

Despite the many difficulties and problems on the road towards inter-Korean economic co-operation, the mandate of the co-operation policy seems to be understood by North and South Korea alike. Although North Korea's system is generally considered to be more vulnerable to alien shocks than the system in the South, it cannot be overlooked that Pyongyang has made prodigious efforts to ensure that it could absorb and digest impacts from China's modernization line and the Soviet Union's *perestroika* policy without a considerable internal shake-up.

To consider North Korea's position today and in the future, North Korea's 'open-door policies' and their prospects need to receive some attention.

North Korea's open-door policies

North Korea's recent open-door policies are closely connected with its joint venture business and changing economic policies. After the enactment of the Joint Venture Law in September 1984,

Pyongyang has displayed a hitherto unseen flexibility in its external policies relating to Western countries. The DPRK denies that it has followed the People's Republic of China in drafting a Joint Venture Law intended to attract foreign investment. However, it is hard to overlook similarities to the Chinese law. It was not entirely unrelated that Secretary Kim Jong-Il and other officials had visited the Shenzhen Special Economic Zone in southern China before establishing its Joint Venture Law.

There is, however, an important difference between the DPRK's open-door policies and China's opening up to the outside world. The Chinese stance has been backed by policies designed to activate the domestic economy by expanding the market price system. Generally speaking, Chinese economic reform policies, including the agricultural contract system and industrial sector autonomy, have been developed gradually in a form that has created momentum on the external liberalization policies.[30] Like two wheels on a cart, the inner-directed market reform policies and the outer-directed open-door policies moved forward together to reform society until the temporary halt following the upheaval in Tiananmen Square in June 1989.

Pyongyang, which continues to place a heavy emphasis on ideology,[31] is not yet moving towards any overhaul or reform of the domestic economy, and this is handicapping trade with the Western world, although the need to attract foreign investment is now recognized. However, Pyongyang has recently tried to break out of its structural stagnation by a series of, admittedly limited, economic reforms. Examples of these reforms are the expansion of farmers' markets, direct sales stores, the recent permission for working people to buy goods at the Nakwon (Paradise) stores in urban areas, the introduction of a productivity-linked payment system in the production of consumer goods, an autonomous business management system in factories, and a material incentives system.[32]

It is also noteworthy that North Korea's third Seven-Year Economic Plan (1987–93) stresses the increase in production of both consumer goods and daily necessities. The plan entailed the creation of a Three-Year Plan for Light Industry (June 1989) and the establishment of a Ministry of Local Industry (July 1989).

While North Korea is currently concentrating its efforts on the fulfilment of its ongoing third Seven-Year Plan, it is placing a large amount of investment into light industries and tourism.[33] Indeed, a vigorous drive is under way to produce so-called '3 August [1984] consumer goods' to implement the policy of the Workers' Party of (North) Korea to start a revolution in light industry.[34] President Kim Il-Sung, in his annual New Year Message of 1990, strongly called for the development of light industries, stressing the need to quickly complete the construction of the Sunchon Vinalon complex, which reportedly will produce annually 1 million tons of carbide, 750 000 tons of methanol, 100 000 tons of vinalon (a synthetic fibre made of limestone), 900 000 tons of nitrogenous fertilizer, 250 000 tons of vinyl chloride, 250 000 tons of caustic soda, 400 000 tons of sodium carbonate and various other chemicals, as well as 300 000 tons of protein fodder.

The main cause of shortages of daily commodities in North Korea lies in the fact that Pyongyang has so far relied on small-scale local industries and on various inefficient work-teams organized in every factory and village for commodity production, while giving top priority to the heavy and defence industries. It is clear that North Korea's recent emphasis on light industries and its flexible attitude towards a pragmatic open-door policy—even though it seems too soon to confirm—signals that a mini-*perestroika* is occurring in the North. When this is added to the foreign joint ventures and the growing tourism industry, a considerable portion of North Korea's economy is seen to be set on a road to quasi-market principles at least.

How much of an about-face Pyongyang can engineer in light of the hitherto raised obstacles to reform remains to be seen. It may depend upon future developments of political relations, both within North Korea's power structure and between the two Koreas. The big swing to, or interaction between, Marxist thought and Friedrich von Hayek's free thought is perhaps not so much an economic as a political issue in the initial stages, with later stages becoming more closely related to economics.[35]

Recommendations for economic co-operation between North and South

In spite of the many differences in ideologies and socio-economic systems between the two Koreas, inter-Korean economic co-operation will not only benefit the two countries economically, but also will help to establish a foundation of pragmatic trust for national unification, not to mention political stability in the East Asian region.

In theory, inter-Korean trade must be mutually beneficial, given the complementarity in natural resource endowments as well as other economic attributes; furthermore, the cost advantages from inter-country transactions cannot be underestimated. Discoordinating factors are the differing political and economic systems of the two countries and their specific internal socio-political conditions. Thus, the extent of future progress in economic co-operation between North and South Korea will depend upon the extent to which the two sides recognize and respond pragmatically to the economic and non-economic benefits and costs involved.

On the ideological and political dimension, both sides can minimize the costs of interchange if they agree to de-politicize reunification issues as much as possible. In addition, both sides should give up the idea that unification must be achieved on one side's terms. The majority of people should have a right to decide what form of government they want to have when the time is right, after a sufficient length of time and experience has been spent on various exchanges between the two countries.

On the social and cultural dimension, North and South Korea need to remember that they share the same culture, history, and language regardless of the alien ideologies under which they have been living for more than 40 years. The continuing dialogues and cultural and economic exchanges between the countries will mitigate the chasm as time goes on.

In order to overcome all existing problems, inter-Korean economic co-operation is badly needed, not least for its practical economic benefits. To promote inter-Korean trade, the following approaches seem to be in order.

First, inter-Korean trade must be treated as inter-country rather than 'international' trade when commodities and services move across the 38th Parallel. In addition, trade and economic exchanges must proceed gradually so as to move slowly towards economic union.

Second, each side should refrain from overwhelming the other by demonstrating any superiority of economic strength so as to cause the other to lose face. All transactions need to be conducted in mutually respectable and face-saving ways. This does not imply that one country should not provide economic assistance of any kind to the other when such help is needed. But even such extensions of assistance need to be arranged so as not to tarnish the image of the recipient side. It might be desirable for both sides to set up joint firms in third countries, which can be used as windows through which flows of capital loan, technology, and other economic inputs are made between North and South Koreans whenever necessity occurs. Of course, it is preferable that such joint ventures should make commercial sense, under conditions that are agreeable to both sides.

For example, the tourism development programme of North Korea could gain momentum in partnership with South Korean businesses. In turn, South Korea may want to open the door for North Korean enterprises seeking joint ventures in South Korea for tourism business aimed at capitalist countries. It is also regarded as a good signal that North Korea hopes to open an air route linking Beijing and Tokyo via Seoul and Pyongyang.[36] In addition, Pyongyang has recently added many tourist facilities, including Western-style hotels apparently aimed at expanding tourism in North Korea.

There is also a feasibility for initiating a free zone area, along with opening the existing demilitarized zone (DMZ) to unfettered communication and travel between North and South. Indeed, there are hundreds of potential projects on which both sides could coordinate their efforts if they were to adopt a positive attitude and a determination to work together. The question is not really 'how' to co-operate and on 'what', but 'if' both sides are truly willing to co-operate in the transition to reunification. For North and South Korean leaders and citizens as well, patience and a genuine willingness to compromise seem to be the most important ingredients for *râpprochement*.

6.6   CONCLUDING REMARKS

The gap between North and South Korea is so great that any dramatic *râpprochement* is unlikely in the short run. Nevertheless, changing internal and external environments are putting enormous pressures on both Koreas to open up to each other. There are however differences in approaches to reunification as well as to issues of economic, cultural, and other exchanges between North and South.

North Korea advocates grand sweeping proposals that focus on big political and military issues. DPRK officials want reunification first, then peace, while the ROK advocates peace first, and reunification later. In October 1980, Kim Il-Sung proposed the establishment of the 'Confederal Republic of Koryo', an idea basically still on his agenda. North Korean leaders want, first of all, to do the following: (1) remove all US forces from South Korea; (2) stage massive joint public events such as music festivals; (3) convene a large meeting of social and political leaders of the two sides; (4) hold tripartite talks with North Korea, South Korea, and the United States; (5) oppose the separate entry application for UN membership of the two Koreas in favour of 'one single-seat joint entry formula'.

South Korea, on the other hand, favours a gradual step-by-step approach, starting with socio-cultural and economic exchanges and only later tackling the larger political and military issues.[37] An example of South Korea's idea of economic exchanges is inter-Korean trade, which would be considered domestic trade. Such trade would involve the exchange of services including the opening up of communications (via postal and telephone services) as well as goods, and these exchanges would help to promote common economic and social values.[38]

Despite North Korea's preoccupation with the political and military aspects of the proposed improvement in inter-Korean relations, there exists an indication of a narrowing gap between the two Korean positions. The countries have just agreed to continue inter-Korean high-level talks which are expected to progress towards a meeting between the two presidents themselves. There also exists a need for initiating bilateral economic co-operation. If economic relations between North and South Korea develop over time, the exchange of products and production inputs will follow, along with capital and technological co-operation. In due course, gradually both sides can adjust the differences between their economic and political systems as they work their way towards national unity. The inter-country trade and economic exchange must be accompanied by the signing of formal economic agreements, with a liquidation accounts system established either between the central banks of North and South Korea or at a chosen bank in a third country. For this, the two Germanys' settlement arrangements may provide the Koreas with useful experiences and guidelines.

In sum, there is no question about whether North and South Korea could make substantial gains through mutual economic exchanges and co-operation in the light of the complementarity of their natural resources and industrial structures. In passing, it is also hoped that the increasing relations of South Korea with communist countries such as the CIS (former Soviet Union), China, Mongolia, and Eastern European countries,[39] combined with the improvement of relations between Japan/the United States and North Korea, will help to improve inter-Korean relations. These developments deserve approval and welcome. The future of progress in economic co-operation between North and South Korea is certainly in the hands of Koreans themselves and will depend on how the two sides recognize the existence of each other and respond to the necessity of co-operation for unification.

In concluding this chapter, two remarks by President Kim Il-Sung of North Korea and President Roh Tae-Woo of South Korea are worth quoting to highlight the future milestones on inter-Korean relations:

Holding the North–South high-level talks is a good thing which allows us to look forward to the future of our people, and co-operation and exchange can be smoothly realized only when the state of political and military confrontation between the North and the South is removed through talks. I hope that by properly advancing the high-level talks, I may consider meeting President Roh Tae-Woo at an early date as possible. (President Kim Il-Sung during meeting with the South Korean delegation to the second Inter-Korean Premier's Talks, Pyongyang, 17 October 1990)

It is most important to build the framework of mutual co-operation through the summit. A South–North summit should be held at the earliest possible date to effectively resolve problems between the two sides.

There is no reason for the South and the North to remain in confrontation at a time when the United States and the Soviet Union, who gave birth to the Cold War, are creating a new world atmosphere of reconciliation. . . .

What is most urgent at this time is to build confidence between South and North Korea. . . . Without exchanges, neither understanding nor confidence comes into being. . . . There were many differences, but the two sides converged on some points. We will be striving to understand the stance of the North to the greatest possible extent, and to accept the North's proposals, if reasonable enough to be translated into action. (President Roh Tae-Woo's selected remarks made during meeting with North Korean Delegation to the first Inter-Korean Premiers' Talks, Seoul, 6 September 1990)

Korean people in both the North and the South, and the outside world as well, will watch to see how faithfully the two political leaders attempt to keep their word. Surely the Korean lamp is 'waiting to be lighted once again for illumination in the East'.

NOTES TO CHAPTER 6

1. See Colin Clark, *The Conditions of Economic Progress*, 3rd edn. (New York: Macmillan, 1957).
2. Based on this 'spirit and method', North Korea was able to mobilize the mass energies of the population, including the armed forces and students, in land reclamation projects. Thus, North Korea's irrigation system is fairly well developed and its grain production continues to increase through the cooperative farming system.
3. Kim Il-Sung in his Fifth Congress report pointed to shortages in fuel, power, and raw materials as obstacles to the full-capacity operation of the industrial plant. He called for a cutback on new construction until the supply of these basic requirements had been augmented. See Youn-Soo Kim (ed.), *The Economy of the Korean Democratic People's Republic* (Kiel: German–Korean Studies Group, 1979), 91.
4. Pyongyang recently announced that the rolled steel output in the iron and steel works across the country jumped 12% in a recent month above the same period in 1989. The Kim Chaek iron and steel complex boosted the output of pig iron and steel 17 and 15% respectively and rolled steel 5%. The Hwanghae iron and steel complex increased pig iron 13% and rolled steel 40% over 1989 by remodelling equipment of processing. The Chollima steel complex increased its rolled steel production by 20% (Pyongyang KCNA, 8 Nov. 1990; and see FBIS-EAS-90–217 (8 November 1990), 18).
5. *Rodong Sinmun* (15 November 1990), 'Let Us Effect Greater Upsurges in Producing People's Consumer Goods' (editorial).
6. *Rodong Sinmun* (3 August 1990), 'Abundant Results, Proud Successes' (article by Hung-Suk Myong).
7. Ibid. Also see FBIS-EAS-90–215 (6 November 1990), 20.
8. Ralph N. Clough, *Embattled Korea: The Rivalry for International Support* (Boulder, Colo.: Westview Press, 1987), 111.
9. The programme included: (1) withdrawal of US forces; (2) subsequent reduction of forces of the two Koreas to 100 000 or less; (3) abrogation of the US–ROK security treaty and the ROK–Japan normalization treaty; (4) guarantee of complete freedom for all political parties and individuals in South Korea and release of political prisoners there; (5) free, democratic elections throughout Korea to establish a unified central government; (6) if South Korea cannot agree to elections, the establishment of a North–South confederation as a transitional step to promote co-operation and interchange between the two Koreas; (7) if South Korea cannot accept a confederation, the establishment of an economic committee to promote co-operation and interchange; (8) a political consultative meeting of all political parties, organizations, and patriotic individuals to discuss all the foregoing matters. See Byung-Chul Koh, 'Unification Policies and North–South Relations', in R. A. Scalapino and J. Y. Kim, *North Korea Today* (New York: Praeger, 1963), 279.
10. The contents of the North–South first joint statement are: (1) that unification shall be achieved through independent efforts, without external imposition or

interference; (2) that unification shall be achieved through peaceful means, not the use of force; (3) that a great national unity, transcending differences in ideas, ideologies, and systems shall be sought first. See Clough, *Embattled Korea*, 113.

11. These countries include Romania, Czechoslovakia, Bulgaria, the Soviet Union, Mongolia, China, Poland, Yugoslavia, and Hungary.

12. Kim Il-Sung, New Year Address (1 January 1990): 'The Concrete Wall Must Be Removed and Free Travel between North and South and a Full-scale Open Door Materialized' (Pyongyang).

13. Roh Tae-Woo, Special Announcement (20 July 1990): 'Steps for Grand Inter-Korean Exchanges of People' (Seoul).

14. However, it is noteworthy that the North Korean Workers' Party newspaper, *Rodong Sinmun*, claimed on 28 October 1990 that the German model of reunification (i.e. the absorption of communist East Germany by capitalist West Germany) could not be applied to the Korean peninsula.

15. Requoted from Peter Murrel, *The Nature of Socialist Economics: Lessons from Eastern European Foreign Trade* (Princeton, NJ: Princeton University Press, 1990), 44.

16. Ibid. 56.

17. Theoretically, there can be completely free-market economic systems in which there is no planning whatever, neither of the physical volume of production nor of prices and wages. However, many market-oriented economies use various degrees of indicative planning today to encourage the self-adjusting processes of the economy to go in a desirable direction.

18. Murrel, *Socialist Economics*, 59.

19. Ibid.

20. Ronald D. Asmus, 'A United Germany', *Foreign Affairs* (Spring 1990), 64.

21. Ibid. 66.

22. Note that South Korea has had a market-oriented but highly government-controlled economy, while West Germany has a 'mixed economy'—part free-market and part socialist.

23. Willi Leibfritz, 'Economic Consequences of German Unification', *Business Economics*, 25: 4 (1990), 7. See also Peter Bofinger, 'The German Monetary Unification (GMU): Converting Marks to D-Marks', *Federal Reserve Bank of St Louis Bulletin*, 72: 4 (1989), 17–35.

24. North Korea recently claimed that the German model of reunification could not be applied to the Korean peninsula (see n. 14). Also, Pyongyang Domestic Service, 1150 GMT, 11 November 1990, asserted that 'reunification of our country should never be such that any one side absorbs and integrates the other side' (see FBIS-EAS-90–219, (13 November 1990), 25–6).

25. In particular, the North's *Rodong Sinmun*, in a commentary on 27 October 1990, declared that 'compromise with capitalism will only bring about destruction'. Therefore, it said that 'North Korea will stick to socialism no matter what changes may take place in the world situation.' In addition, it is noteworthy that to date the *rȃpprochement* between North and South has been over-shadowed by the South's bias towards the improvements of relations with the other communist countries.

26. North Korean Central Broadcasting Station (KCBS), 25 October 1990. See also *North Korea News* (published weekly by Naewoe Press, Seoul), no. 551 (5 November 1990), 4 and 5.

27. Elizabeth Pond, *After the Wall: American Policy toward Germany*, a Twentieth Century Fund Paper (New York: Priority Press Publications, 1990), 10.

28. Ibid. 10.

29. The Rangoon bombing of 1983, a North Korean attempt to assassinate President Chun Doo-Hwan visiting Burma, killed seven high-ranking South Korean officials. Also, North Korean agents caused an explosion aboard a Korean Airline plane in November 1987, killing all 115 passengers and crew. This explosion was interpreted as an attempt to disrupt the 1988 Olympic Games at Seoul.

30. Masao Okonogi (ed.), *North Korea at the Crossroads* (Tokyo: Japan Institute of International Affairs, 1988), 158.

31. North Korea has recently been strengthening its emphasis on the importance of ideological revolution as well as ideological education. See *Rodong Sinmun* (5 December 1990), 'Ideological Revolution is Most Positive Method of Ideological Remoulding'; also *Rodong Sinmun* (14 July 1990), editorial on the North's intensified indoctrination campaigns.

32. Ha-Cheong Yeon, 'Bridging the Chasm: Co-operative Economic Relations between South and North Korea', Korea Development Institute Working Paper no. 9012 (Seoul: KDI, August 1990), 8.

33. See *Rodong Sinmun* (14 November 1990), editorial; also *Rodong Sinmun* (3 August 1990), 'Abundant Results, Proud Success' (by Myong Hung-Suk); *Kulloja* (Pyongyang), no. 8 (August 1990), 67–72: 'Improving Commercial Service is an Important Task to Guarantee the People's Conveniences in their Lives' (article by Han Chang-Kun). North Korea is now constructing many tourist facilities to attract foreign visitors and placing tourism advertisements in a Chinese newspaper, *People's Daily*. Pyongyang also opened its first golf course of a total area of 60 000 m² on 18 August 1990.

34. KCNA (Pyongyang), 6 June 1990. See also FBIS-EAS-90–111 (8 June 1990), 13. The 3 August (1984) programme was launched by Kim Chong-Il for the promotion of consumer goods production. This production is organized locally at home and workplace to produce consumer goods by utilizing industrial by-products and waste materials. The products are sold directly to district consumers through what are called 'direct sales stores'.

35. By this I mean that the two Koreas' co-operative relations must first be established by ongoing high-level political talks, and then inter-Korean economic co-operation can smoothly proceed.

36. According to the North Korean Central News Agency (KCNA) report on 7 August 1990, North Korean Deputy Foreign Minister Kim Yong-Nam said, during talks with visiting International Civil Aviation Organization (ICAO) officials, that North Korea hopes an air route between Beijing and Tokyo via Seoul and Pyongyang will be opened soon. See also *Tong-A Ilbo* (Seoul), (14 October 1990), 'Seoul–Pyongyang Air Route to be Discussed in Montreal, Canada, between North and South Korea during October 22–26'.

37. The South Korean government unilaterally announced in June 1989 that it would permit inter-Korean academic and cultural exchanges. Since then, 181 programmes had obtained government approval by 4 December 1990. Out of the 181 approvals, only 32 (18%) have been realized at time of writing, mostly in third countries. The exception was the music event held in September 1990 in Pyongyang, with a score of southern musicians participating. See *Korea Times* (Seoul) (4 December 1990), 3. Also noteworthy is the fact that some South Korean professors have recently pressed for bolder North–South exchanges. At a seminar held recently in Seoul, professors requested that 'statesmanship is needed on the part of the South to try to induce even politically hostile groups to become favourable to us, if the South is to open itself and also is to seek to have the North open its doors, and if there is to be no problem in security environments'. They also warned that, 'if the South government seeks

to intervene in the inter-Korean student contacts in a more-than-needed authoritarian manner out of distrust of the North, it will make South–North student exchanges all the more difficult'. See *Korea Herald* (4 December 1990), 3.

38. Kong Dan-Oh, 'Intra-Korean Relations and the Strategy of Rapprochement', unpublished paper presented at the 'Conference on Korea in the 1990s', Pomona, Cal. (24 February 1990), 8.

39. Two-way trade between South Korea and communist countries, boosted by the South's 'northern policy', will likely reach $8.2 billion in 1991, up 70% from 1990's projected $4.8 billion. During the first seven months of 1990, South Korea exported $229 million worth of goods to the Soviet Union and imported commodities worth $187 million. During the first eight months of 1990, South Korea imported $110 million worth of goods from East European countries and exported $303 million for a two-way trade of $413 million. See *Korea Herald* (21 October 1990), 6.

7

Relations with Other North-East Asian Countries

Social dilemmas occur when outcomes that are good for each group member acting individually are bad for the group as a whole. In the language of game theory, they occur when payoffs to each participant yield dominating decision strategies that converge on a deficient equilibrium.

'Cooperation for the Benefit of Us—Not Me, or My Conscience', by Robyn M. Dawes, Alphons J. C. van de Kragt, and John M. Orbell[1]

7.1 INTRODUCTION

As we head towards the twenty-first century, the world is changing so rapidly that any rational forecast of even near future events is simply not feasible. Today, nations are reshuffling their relations with neighbours as the old environment of the Cold War era is being replaced by the emergence of a new economic regionalism and its interaction of economic incentives. Are these to be interpreted as 'essential changes'? The answer may be either 'yes' or 'no', depending upon our point of view. It could be 'yes' if we look at things from the local perspective within the finite time horizon, but 'no' from the universal perspective of infinitely repeating history. As Su Tung-Pao, a renowned writer of the ancient Sung Dynasty of China, said, 'When viewed with a belief that things are bound to change, there is not even a single element in the heaven and earth that does not change, while all materials in this world and the existence of myself who lives and dies are eternal when viewed from the standpoint that things do not change.'[2]

Really, everything could be interpreted as being in a constant process of change when we perceive the world from the changing aspects of time, while even in a flower that blooms in the morning and withers in the evening we could find something endless and eternal since the period of its recurrent blooming stretches from ancient time to the endless future.[3]

The hard struggle of former command economies to transform themselves into market economies has been taking place ever since the late 1980s with the fascinating diffusion of Mikhail Gorbachev's ideas of *perestroika* and *glasnost* in the former Soviet Union, as well as in Eastern Europe and China. This wind will not leave North Korea untouched for ever, even though it appears not yet to have disturbed that hermit kingdom. When it hit East Germany the resulting disruption led to the unexpected German reunification in 1990, which was followed by many equally unexpected events across the whole of Eastern Europe and the Soviet Union as well. The Soviet Union itself dissolved after the abortive coup in August 1991. As William Nordhaus pointed out,[4] the communist countries, having spent up to seven decades systematically attempting to establish command economies, have executed an abrupt about-face in an attempt to create political and economic institutions in the style of market economies. Were Karl Marx and Friedrich Engels alive today, they could not help grieving for the fading Marxist ideal the world over.

The new world order and the changing environment enveloping the Korean peninsula have caused the two Koreas to adopt a somewhat more practical approach to the issue of their relationship. The ever intransigent North Korean leadership reversed its tack and sought simultaneous entry into the United Nations with South Korea on 17 September 1991. Thus, both North and South Korea became members of the UN family at the same time, effectively transposing their battleground from a desolate outer plain to the inner ring of the United Nations. This should make the division considerably easier to deal with. Furthermore, there seems to be a general consensus among Koreans that the demand for mutual exchanges and economic co-operation between North and South is even greater than ever.

It is true that most Koreans on both sides of the divide would like to see the two halves ultimately reunited in some way. Although politicians and laypeople alike understand that today's realities do not allow for such a reunification in the near future, they nevertheless believe that, even before such time is right, there are genuine prospects for economic collaboration between the two Koreas, while meanwhile maintaining the status quo. Indeed, limited economic exchanges have been occurring off and on ever since October 1988. While it is true that economic collaboration between the North and the South still encounters a range of obstacles as a result of the wide differences in both political and social values, this economic link has been growing openly ever since both parties signed the Basic Agreement on Reconciliation, Non-aggression, Exchanges and Co-operation on 13 December 1991. The Basic Agreement called for the suspension of mutual slander and destructive activities, for non-interference in domestic affairs, and for the promotion of inter-personal and economic exchanges. In spite of the differences of covert intention between the two sides regarding this agreement, and the confrontation over the question regarding practical steps towards the substantial implementation of the accord, both parties seem at least to be sincerely trying to derive some tangible result from the continuing series of the prime-ministerial talks that began in September 1990.

At the end of the seventh round of these talks, held on 6–7 May 1992 in Seoul, both sides agreed to work out a plan to allow elderly persons to visit family members in the other half of the divided country. At working-level contacts on 12 June 1992, it was agreed to exchange a 240-member delegation from each side, consisting of 100 elderly family visitors, 70 performing artists, and 70 others including journalists and support personnel. This programme, scheduled to take place on 25–8 August 1992, would have been the first such event since September 1985, when visiting groups, consisting of 151 people from each side, visited the other's capital city for brief family reunions. However, the inter-Korean family reunion programme, as well as Seoul's proposal for free choice of residence,[5] has been disbanded because North Korea has again stipulated that South Korea should drop its insistence on nuclear inspection and return aged communist Lee In-Mo[6] to the North before procedural matters can be decided relating to the cross-border visit programme.

Seen in this perspective, it is clear that non-economic talks are still confined mostly to procedural issues for further discussion, and the dialogue remains marked by largely incompatible negotiating styles and objectives.[7] But package negotiations of a set of political, economic, and military issues between the two countries are likely to improve in the near future, in light of the recent trend of economic co-operation among neighbouring countries through the provision of free and indiscriminate access to each other's markets. This movement began to gain momentum in the early 1990s, not only with the collapse of the former Soviet Union but also with the enlarging economic regionalism including the European Community (EC), the European Free Trade Area (EFTA), and the North American Free Trade Area (NAFTA). Against such strengthening of economic regionalism in Europe and North America, countries in north-east Asia have begun to recognize the need to explore the possibilities of regional economic co-operation among themselves. Thus, the inter-Korean economic transaction can be viewed not only from the angle of the inter-Korean economic situation and objectives, but also from the broader perspective of north-east Asian economic development.

The framework of north-east Asian regional co-operation would receive a very significant stimulus to political and economic development and stability if the relations of the two Koreas were improved and their roles in the area were well co-ordinated. The increasing dialogue between the two countries and the growing degree of economic collaboration may be expected to enhance the opportunities for co-operative effort among all countries in the region, including China, the Koreas, Mongolia, Russia, and Japan. Of course, the extent to which the economic links will contribute to fundamental changes in the North–South Korean relationship remains to be seen, because such economic partnership is now being pursued against the context of continuing political rivalry and military confrontation. Nevertheless, a good economic relationship between North and South would serve gradually to reduce the level of tension, whatever the state of the other dimensions of the relationship. The two Koreas have openly expressed their interest and support for the development of Special Economic Zones (SEZs) in the north-east Asian Region. In particular, the Tumen River project, proposed by North Korea under UN Development Programme sponsorship in October 1991, has attracted a good deal of interest in both North and South Korea.

In this chapter four major areas of concern will be considered. First, the prospect of North Korean economic policy changes will be assessed. Second, in view of the changing international environment, the shifting relationships among North and South Korea, China, Russia, Japan, and the United States will be discussed, from which it may be possible to diagnose the development of a possible new patronage of the two Koreas which could, over the medium to longer term, result in worth-while economic returns for both parties. Third, the evolving environment for inter-Korean transactions in the framework of north-east Asian economic co-operation will be considered. Fourth, the region's defence and security implications in relation to North and South Korea will be briefly reviewed. Lastly, the outlook on Korea's reunification towards the year 2000 deserves some discussion along with a rough estimate of the cost of reunification.

7.2 PROSPECTS FOR EXTERNAL ECONOMIC POLICY REFORM IN NORTH KOREA

North Korea's external economic policy must be functionally re-lated to its internal economic and political situation as well as to its political policies concerning its neighbours in these times of change. Until recently, however, North Korea has maintained a system of totalitarian autarky under the repressive system imposed by President Kim Il-Sung, as described in detail in earlier chapters.

In the mid-1980s, North Korea first attempted to open up its economy by seeking joint economic projects with foreign enter-prises, but these had little effect on the economy other than to increase foreign debts as imports relevant to these projects rose sharply. To make matters worse, the bombing of Korean Air Line (KAL) 858 with all 115 persons on board in November 1987 by North Korean agent Kim Hyon-Hui and her co-perpetrator tended to make potential Western investors shy away from North Korea's joint projects. Thus, following the sudden economic collapse of its most important trading partner—the Soviet Union—and because of declining economic support from China, the Pyongyang regime found itself under mounting pressure to seek co-operation, even with such old foes as South Korea, Japan, and the United States.

While preparing to seek practical ways to adapt itself to changes in the world situation from early 1991, North Korea signalled its interest in establishing a Special Economic Zone development project on the lower portion of the Tumen River. It revealed that it was seriously examining the idea of the joint development of this area during meetings sponsored by the UN Development Programme (UNDP): the first in Ulan Bator, Mongolia, in July

1991, and the second in Changchun, China, on 29–31 August 1991. The Changchun seminar on north-east Asian economic co-operation, to which more than 60 scholars and officials from six countries (North and South Korea, Japan, the Soviet Union, Mongolia, and the host country, China) were invited, was organized jointly by the Chinese Asia Pacific Research Institute and the East–West Center in Hawaii under the auspices of the UNDP.

The proposal for a Tumen River development project had originally been initiated by China a year earlier, and China indicated that she wanted to set up a free trade zone in the city of Hunchun. In the meantime, at a third round of UNDP-sponsored meetings held in Pyongyang on 18 October 1991, North Korea proposed to develop a Special Economic Area around Sonbong (formerly Unggi) in the north-easternmost section of the Korean peninsula. To date, North Korea has started preliminary business talks with enterprises from several countries, especially concerning the construction of fisheries, clothing manufacturing, electronic products assembly plants, and so on. How the proposed plans will be co-ordinated and finalized among both the countries involved and the UNDP remains to be seen. After two or three years of feasibility studies, the UNDP is expected to present a final compromise plan if differences among North Korea, China, and Russia over where and how to proceed fail to narrow.

It is worth mentioning that, while North Korea is cautiously signalling a hidden opening of its economic policy, the North Korean Workers' Party has repeatedly been urging the people to 'believe firmly in the superiority of our socialist economic system based on the *chuche* ideology'.[8] In spite of the country's extreme sensitivity about the outside world's expectations of 'possible overall changes' in the North, moderation in the North Korean economic policy appears inevitably bound to come about very gradually, and only because it is not possible for such a small economy to sustain an isolated self-sufficiency policy in the highly inter-dependent world. While North Korea's objective economic needs necessitate its opening up to the outside world, the political risk of exposing its people to the visibly superior South Korean economy may be deemed too great for Pyongyang to take the plunge for the moment. This and other signs suggest uncertainty, and no doubt controversy, in Pyongyang, but it is very highly probable that the North's technocrats favour an eventual policy of relaxation.

What pattern will North Korea follow if it is to reform its economic policy? Several models appear in order,[9] but the most relevant one seems to be the China model, which could be described as a *perestroika* without *glasnost* in its early stages. China adopted its economic reform and open-door policy in 1978 after having experimented with the tightly closed Stalinist system ever since 1949. It was about 20 years behind Yugoslavia, which initiated its reforms as early as the beginning of the 1950s, or Hungary and Czechoslovakia, which took up similar reform policies in the mid-1960s. The differences between the reform policies of China and Eastern European countries were that China introduced a responsible management and production system into the rural sector *before* the urban industrial sector, while most Eastern European countries focused on the reform of industrial management and state enterprises. In addition to centring its reform policy mainly on agricultural management and the incentive system, China dared to introduce its external open-door policy in 1979 with an eye to narrowing its gap with other developing countries. In particular, China has developed five Special Economic Zones to induce foreign capital and technology since 1979.[10] By the end of 1986, China was able to induce more than 4700 foreign enterprises to sign for joint ventures in the country with about US $220 billion of investment from abroad. The Chinese economic reform process has been stalled for a couple of years as a result of the Tiananmen Square massacre in June 1989. However, China is now back on course for its planned economic reform, which in turn is naturally accompanied by a considerable extent of political reform, seen or unseen.

As explained earlier, North Korea has recently encountered a drastic erosion of its international position. With the loss of Soviet patronage, it seems to be attempting not only to maintain Chinese political and military support, but also to develop ties in the Asia–Pacific region as well as contacts with Japan and the United States.

As far as seeking the development of contacts with market economies is concerned, it should be more or less self-evident that the Pyongyang leadership is aware of pressures to bring about substantial economic and political change. To minimize any setbacks in the legacy of its monolithic power structure, therefore, North Korea is likely to choose gingerly to try something like the pattern of reform that China has implemented. In other words, it will have to agree to follow an early-stage pattern of the Chinese economic reform programme, though in a somewhat limited scope, while maintaining the coercive social order which still seems to be immune from substantial challenge, both internal and external. President Kim Il-Sung's visit to China during 4–15 October 1991 included a tour to both industrial and farm facilities in the Jinan, Shantung, and Jiangsu provinces. His visit was interpreted as having undoubtedly provided some inducement for Pyongyang to seek changes in its economic policies.

In any case, it cannot be denied that the North Korean leadership could face a crisis if its social and political system remains unaffected by the spill-over effects of economic change.

With limited access to regular sources of foreign capital, at least for now, North Korea cannot help but seek joint ventures aimed specifically at the export market in three main ways: (1) through international organizations such as UNDP and UNIDO, (2) from Korean residents, largely in Japan and the United States, and (3) through economic co-operation with enterprises in South Korea. However, unless North Korea dares to initiate changes in its economic policy, the probability of its securing any large-scale investment programmes is likely to remain insignificant. Of course, the options available following an opening up of its economic policy would be likely to endanger the country's political and social stability over the medium to longer term. That is why the benefits to be derived from economic policy reform need to be weighed very carefully. In turn, it is this factor that constitutes the essential complexity in the Korean peninsula, with both countries being exposed to highly sensitive economic, political, and military issues.

7.3   THE DEVELOPMENT OF INTER-STATE RELATIONS IN THE NORTH-EAST ASIAN REGION

A strengthening of economic relationships is creating new order and dimension among the countries of the north-east Asian region. The opening of diplomatic and trade ties between South Korea and the CIS (former Soviet Union), China, and Mongolia will sooner or later entail similar relationships between North Korea, Japan, and the United States, not to mention hopes for an improvement in inter-Korean relations.

South Korea has pursued full-scale efforts to develop economic and political *rậpprochement* with its neighbours, grouping all economic and diplomatic contacts and negotiations with Russia, Mongolia, China, and North Korea under its established 'northern policy' umbrella. Indeed, South Korean successes in establishing official diplomatic relations with the USSR on 30 September 1990, and with other Eastern European countries, have led on to its trade relationship with China and the development of the inter-Korean dialogue.

7.3.1   *South Korea's Relations with China and the Soviet Union*

The relation between any two countries is interlaced with the national interests of each nation involved. As long as there exists some chance of mutual benefit in either economic or political terms, economic and/or political interaction can grow between countries. In the well-known language of Paretian welfare economics, if pay-offs to one party are increased, and yet no one is worse off, interaction is increased.

The national interest (or pay-offs) can be viewed from two dimensions: economic and political. The Russian and Chinese approaches to South Korea have perhaps been motivated more or less with a view to gaining access to South Korean capital and technology than with political and diplomatic objectives. On the other hand, South Korean policy-makers have apparently placed greater emphasis on the security, political, and diplomatic dimensions of the 'northern policy' than on its economic sphere. South Korea's success in developing closer relations with formerly hostile nations such as the Soviet Union and China has enabled the country to take a more positive stance in its efforts to increase contact with North Korea and to seek some form of *rậpprochement*. In turn, the collaboration between the two Koreas is one that would suit the Chinese government extremely well and would ensure that the multilateral economic and political linkages continue to expand.

China has made it clear to the two Koreas and those major powers with interests in the peninsula that Beijing does not wish to see North Korea completely isolated or pushed into a corner. Nevertheless, China acknowledges that its economic interests lie more with South Korea.[11] In particular, since the 1988 Olympics in

Seoul, the economic exchange between China and South Korea has greatly improved. The bilateral trade volume rose from $149 billion in 1987 to $308.7 billion in 1988, $314.3 billion in 1989, and $384.8 billion in 1990. China emerged as the eighth largest trading partner of South Korea at the end of 1990. (In fact, in 1992 South Korea was the fourth largest market for Chinese exports.) Although the sum total of its investment in China is not great, South Korea has made some direct investments in Shandong and Liaoning. The non-government trade between South Korea and China's coastal areas will grow at high speed. Boats carrying passengers and cargo now run between Inchon in South Korea and Woihai and Tientsien in China, and there are regular chartered air flights between Seoul and both Shanghai and Tienjin, making travel and transportation more convenient. It can be expected that any form of *râpprochement* on the Korean peninsula will serve to increase the economic and political ties between South Korea and China. The North Korean leadership is well aware of this unavoidable situation and of the irreversibility of China's policies regarding the peninsula.

The relationship between South Korea and Russia (former Soviet Union) began to expand rapidly from the mid-1980s. The total trade volume between the two countries increased from $110 million in 1985 to $890 million in 1990. The share of the former Soviet Union in South Korean trade with all communist countries changed from 9.5% in 1987 to 19.1% in 1990. At the Roh and Gorbachev summit meeting held in April 1991 in Jejoo Island off the South Korean coastal area, South Korea promised the Soviet Union that it would supply an economic grant of $3.0 billion to the USSR over a three-year period starting in 1991. This economic co-operation fund consists of $1.5 billion for consumer goods and raw materials, $500 million for capital goods, and $1.0 billion for a cash loan.

Not only the South Korean concern to maintain the status quo in the peninsula for the time being, but also its interest in gaining access to the Russian Far Eastern natural resources drive it rigorously to seek closer ties with Russia. On the part of Russia, the closer economic co-operation with South Korea in the Asian Pacific region means new opportunities for the development of access to the external sources of funds and other resources that are badly needed to overcome its economic crises. Of course, in Russia there remains a degree of latent sympathy for North

Korea, and certainly an understanding of its sense of isolation and a concern over its precarious strategic situation. Despite this, however, there seems to be little chance of a revival of the old relationship between North Korea and Russia even though the two countries have a friendship treaty that will remain in force until 1995.

As China's willingness to provide genuine support for North Korea has been reduced in recent years, and as the North Korea–Russia relationship has been put under serious crisis, North Korea is looking to improve its relations with Japan and the United States, while making some effort to improve the North–South relationship, which would be viewed favourably by China, Russia, the United States, and Japan to a lesser extent.[12]

7.3.2   North Korea's Relations with Japan and the United States

North Korea's renewed efforts to develop its relations with Japan have been made rigorously since 1990, but without much progress in spite of Kanemaru's support for the idea. The Japanese government is taking a strong position to back the US policy that links economic and political contact with North Korea to North Korea's compliance with the inspection provisions of the nuclear safeguard agreement. Pyongyang has initially responded to this situation by maintaining an inflexible stance in its official talks with the Japanese; and Japan in turn has reaffirmed that it would not normalize relations with North Korea unless the nuclear issue was resolved. This has had the effect of advancing the progress of the inter-Korean economic talks considerably more than if Kanemaru's push for a resumed Japan–North Korean relationship had been supported. Moreover, North Korea, in a stunning move, has now indicated that it will accept special or random inspection of its nuclear facilities by the International Atomic Energy Agency (IAEA), as Richard Solomon, US Assistant Secretary of State for East Asian and Pacific Affairs, reported on 8 July 1992.

Should North Korea actually take the realistic position of opening up its nuclear facilities to outside inspection, the DPRK–US and DPRK–Japan relationships are likely to enter a new era, thus paving the way for economic and political dividends to Pyongyang. In addition to the stalemated nuclear issue, North Korea's debt to

Japanese companies, totalling about $640 million accumulated since the 1970s, has been a barrier to developing trade between North Korea and Japan, which has amounted to only about $500 million a year in recent years. Nevertheless, it is likely that the Japanese will seek to expand their business involvement quickly in joint venture projects and trade as North Korea becomes more open to the world economy—inside accepted parameters.

Of course, the linking of the economies of the two Koreas would impact on and from the various relationships and alliances that concern the north-east Asia region. If such economic collaboration is to be a success, the other countries in the region will be required to contribute in one way or another—for example by demonstrating their support for the development of Special Economic Zones (SEZs) in the region. In particular, the Tumen River project is currently attracting a good deal of interest as a candidate for economic co-operation between North and South Korea, China, Russia, and Japan. North-east Asian economic co-operation, if ever organized,[13] could offer North Korea the chance to improve its economic circumstances without immediately being obliged to compromise its ideological base in any obvious way. Furthermore, North Korea could use Japan or other countries as a source of financial and technological support without fear of opposition from South Korea, if multilateral economic co-operation is promoted in this region.

Finally, it is worth noting that Pyongyang is desperately seeking to increase various levels of contact with the United States. Most recently, in June 1991 and in May 1992, North Korea has returned the remains of a few US servicemen who had been missing in action in the Korean War.

7.4 PROSPECTS FOR CO-OPERATION

A discussion of the workable form or modality of north-east Asian economic co-operation should begin with a definition of the region. North-east Asia, in narrow geographical terms, encompasses North and South Korea, Japan, Mongolia, a part of China (specifically, the eastern and northern provinces), Japan, a part of the former USSR (i.e. the Far East and Siberia within the maritime province). The area covers about 17.5 million square kilometres, that is, 40% of the whole Asian continent, and comprises a population of 580 million, 20% of the total in the continent (see Table 7.1). Most of these countries have had varying economic and political structures and systems over the last couple of decades, although they are relatively homogeneous in their culture, which is rooted more or less in Confucian ideas. Even among the market economies in the area, the individual structures reveal considerable gaps and significant heterogeneity in their levels of development—for example, compare Taiwan, with its commercial–industrial base composed of many small and medium enterprises, with the conglomerates characteristic of Japan and South Korea. Taiwan, South Korea, Hong Kong, and Japan all have market economies, and Japan has been the leader in the market-oriented development strategy in the region. On the other hand, China, Mongolia, Russia, and North Korea have all had command economies, which have without exception suffered progressively from severe structural imbalances, a weak managerial system, poor incentives, low productivity, and faulty planning (see Table 7.2).

A problem, therefore, is that these north-east Asian countries have very differing views on the scope and nature of economic co-operation. China and North Korea tend to view regional co-operation as entailing the participation of the Western countries in regional development projects such as the Hunchun development project (China) and the Tumen River development project (North Korea). By contrast, Japan seems to envisage it in a much broader and longer-term framework of vertical and horizontal division of manpower and other resources among countries in the region. Although there are many barriers and bottlenecks to be overcome before the realization of any effective economic co-operation in north-east Asia, the growing weight of the region in the world economy appears likely to pave the way for greater co-operation among all the constituent countries in the 1990s.

7.4.1 The Tumen River Area Project

One of the most promising possibilities for multilateral economic co-operation in the region has been the UNDP-sponsored Tumen River area development project. Indeed, a North-east Asia Sub-Regional Programme meeting was convened on 6–7 July 1991 by the UN Development Programme in Ulan Bator, Mongolia,

TABLE 7.1 The economic potential of major north-east Asian countries, 1989

	S. Korea	N. Korea	China	USSR	Japan	NEACs[b]	World
Land ('000 km^2)	99	123	9600	2240.2	37.8	3260.2	1 495 000
Population (million)	42.7	21.9	1 102.4	288.7	123.2	1 577.9	5 227.0
Farm population (% of total population)	19.5	41.6	60.2	19.7	7.1	—	—
Mineral and other resources ('000 tons)							
Copper	0.1	12.0	370.0	990.0	16.7	1 388.8	8 791.8
Iron	370	4 000	54 173	148 800	158	207 501	546 993
Steel	12 578	5 900	57 040	114 558	79 221	269 297	537 330
Oil	0	0	136 823	622 130	0	758 953	2 913 018
Coal	24 295	40 000	946 460	599 000	11 223	1 620 978	3 453 970
Rice	6 053	6 267[a]	169 110	287 000	10 499	478 929	—

[a] Total rice production in 1984.
[b] Total north-east Asian countries including Mongolia, Taiwan, Hong Kong, etc.

Source: UNCTAD Commodity Yearbook 1990.

TABLE 7.2 The mutually complementary conditions of north-east Asian countries

Nations/regions	Advantages	Disadvantages
Japan	Capital savings; advanced technology; plenty of superior equipment ready to move out; vanguard industrial products and management experiences	Severe shortage of energy and industrial resources; insufficient grain for animal husbandry and some agricultural products; comparative deficiency of labour
'Soviet' Far East	Plenty of forests, non-ferrous metal ore; aquatic resources; oil, gas, coal, and some products of heavy and chemical industries (such as steel, fertilizers)	Severe shortage of agricultural and light industrial products; lack of labour and capital; backward industrial equipment and management experience
North-east China	Favourable agricultural conditions; adequate and various agricultural products (such as maize, soya beans, meat, fruit); some textile industrial products; oil, coal, building materials; Chinese medicinal herbs; and excess labour	Lack of capital, advanced equipment, technology, and management experience; comparative shortage of some mineral resources; conditioned infrastructure
North Korea	Rich mineral resources, metal ore, and simple processed products; aquatic products; some industrial commodities; and plentiful labour	Shortage of capital; insufficiency of farm, sideline, and light industrial commodities; backward equipment and technology
South Korea	Surplus capital; advanced technology and equipment ready to move out; vanguard industrial products	Shortage of energy and industrial resources; lack of grains for stock-raising; insufficiency of labour
Mongolia	Plentiful products of animal husbandry and of mineral ores, especially fluorspar	No convenient means of communicating directly with other north-east Asian nations; lack of capital, technology, equipment, farm products, and light industrial commodities

Source: Chen Cai, Yuan Shureu, Wang Li, and Ding Sibae, 'Regional co-operation in North-east Asia and the Exploitation of Triangle Area of Lower Tumen River', paper presented at the Second International Conference on the Economic and Technological Development of Northeast Asia, Changchun, 1991; available from the Geographical Institute of Northeast Asia, Northeast Normal University, Changchun, Jilin Province, 130024, People's Republic of China.

involving representatives from China, North Korea, South Korea, and Mongolia. In this meeting, all countries accorded top priority to the Tumen River area development project. The UNDP was requested to assist these participating nations in carrying out the initial feasibility study, selecting strategic locations with exceptional potential, and co-ordinating and implementing the project's initial phases, as well as advising with respect to long-term regional development.

In late August and early September 1991, the UNDP met again with representatives from these countries as well as observers from Japan and Russia at the second North-East Asia Development Conference held in Changchun, Jilin Province, China. A UNDP-sponsored mission then visited the three countries bordering the site—China, Russia, and North Korea—to obtain an on-site view of the area. The mission reported to the participating countries that the strategic location of the Tumen River delta area, stretching from North Korea's Najin port to China's Hunchun to the Russian port of Posyeta, has enormous potential in terms of both resources and global trading patterns. It is an area situated within easy access to major markets in the industrialized Chinese provinces of Jilin and Heilongjiang, and to favourable supply factors such as labour and natural resources from Russia, North Korea, and Mongolia (see Table 7.1). The Tumen River area has the additional advantages of proximity to Japan and South Korea and of providing access to Europe.

There are however differences between the participants in some important respects, as China, Russia, and North Korea each wants to develop one or more areas in its own territory as economic zones. 'The challenge will be to find an approach that can maximize the benefits for all the parties concerned, while also taking into account their respective articulated objectives. This approach might not necessarily adhere exactly to each suggested way of achieving these objectives.'[14]

According to the UNDP mission report, this plan will need to be implemented in a co-ordinated manner in order to: (1) secure the confidence of the international investment community and the necessary finance—about $30 billion for the long-term development including various infrastructure facilities; (2) avoid unnecessary or costly duplication of facilities; and (3) eliminate unhelpful or destructive competition.

7.4.2 The Positions of North and South Korea

The North Korean proposals for north-east Asian economic co-operation are set out in a study report on Rajin–Sonbong economic trade produced by the DPRK's External Economic Commission. In view of its economic backwardness, North Korea's new interest in special Economic Zones is of significance. Not only can this be interpreted as a signal that the DPRK leadership is prepared to countenance controlled involvement in the external economic relations on a greater scale in times to come; it also indicates that it will seek greater economic exchange even with previously hostile capitalist countries, provided the ties supply the North with much-needed foreign capital and advanced technology.

North Korea's leadership seems to be well aware of its economic situation and of the new chance for economic take-off offered by the establishment of such economic zones. Its approach is understood to involve establishing a controlled zone in which export-related industries could be set up as joint ventures using Japanese, South Korean, or other funds. The North Korean government is also taking account of the growing demand in Russia and north-east China for port facilities. South Korea is responding very positively in support of the North Korean approach, while the ideological, political, and military division between the two countries continues to act as a limiting factor in their economic and political co-operation. The on-going dialogue between the two Koreas will continue, as, no doubt, will the state of mistrust and hostility that is characteristic of their relations.

The scope for real collaboration is still limited, but there is ample scope for the development of close political and economic interchange between the two Koreas in the area of multilateral co-operation. On the one hand, both governments are well aware of the need to appear responsive to popular hopes for a better inter-Korean relationship and ultimately for a reunified country. Also, they recognize that greater collaboration along both economic and political lines not only would contribute gradually to easing the age-long state of mistrust and hostility, but also would demonstrate some degree of Korean solidarity whenever an issue such as the selection of a development project is placed on the agenda of the multinational north-east Asian community. Indeed, it is very much hoped that the trend and pattern of inter-Korean co-operation for the Tumen River project and other areas will continue, although the evolution of their respective political agendas could come positively or negatively into play over the short to medium term. The Korean peninsula is the bridge linking the economies of the Pacific and continental north-east Asia. Thus, a tension-free Korean peninsula is a very important factor in promoting economic co-operation in the North Pacific region.

This is not to overlook the fact that a co-operative and harmonious environment in the region can play an even more important role in improving inter-Korean relations.[15] To make this clear, we shall examine the evolving relation of the two Koreas in terms of both bilateral and multilateral co-operation in north-east Asian development in the sections that follow.

7.5   THE CHANGING ENVIRONMENT FOR INTER-KOREAN TRANSACTION IN THE DEVELOPMENT OF NORTH-EAST ASIA

North and South Korea agreed in principle on the implementation of trade and economic co-operation and the establishment of the South–North Joint Economic Co-operation Committee on 20 June 1985 at Panmunjom (see Appendix 1). But until late 1988 there was neither economic exchange, nor any substantial improvement in the inter-Korean relations. At last, on 7 July 1988, the South Korean President Roh Tae-Woo proposed a new effort to begin an era of national self-esteem, reunification, and prosperity by building a social, cultural, economic, and political community in which all Koreans could participate. To that end, he declared that 'we will actively promote the exchange of visits between people from all walks of life in North and South Korea, and will open the doors of trade between the two Koreas which will be regarded as internal trade within the national community'.

The mutual realization of the necessity of inter-Korean exchange and the favourable international environment thus led to the opening of indirect trade in October 1988. Nevertheless, the volume of the economic exchange between North and South has remained minimal. From October 1988 to December 1991, South Korea's imports from the North totalled $136 655 000 with 268 items. In the same period, South Korea exported a total of 28 items amounting to $6 803 000. Following the signing of the basic agreement on reconciliation, non-aggression, and exchanges and co-operation between the two countries on 13 December 1991 (see Appendix 2), inter-Korean trade has been greatly on the increase. During the first half of 1992, inter-Korean merchandise trade amounted to $106 891 000, up 31% over the corresponding period of 1991. Imports of northern goods to the South rose 43%, from $68 592 000 in the first half of 1991 to $98 030 000. On the other hand, shipments of ROK products to the North decreased as much as 31%, from $12 874 000 to $8 861 000.

The main items imported into the North in January–June 1992 were chemicals worth $5 972 000, steel worth $2 102 000, and textiles worth $223 000. DPRK goods that were shipped to the South during the period were agricultural and forestry goods ($8 122 000), fisheries ($13 143 000), steel ($7 820 000), non-ferrous metals ($43 729 000), chemicals ($396 000 000), construction equipment ($22 484 000), and textiles ($1 341 000). In particular, non-ferrous metals, construction equipment, steel, and other raw materials accounted for 76% of the DPRK goods imported into the South in the first half of 1992, with the remaining 24% comprising farm, forestry and fishing goods, textiles, and other consumer products. Other important items shipped from North to South since late 1988 have included zinc ingots, gold and silver nuggets, hot-rolled coil, billets and dried fern, etc. Southern goods shipped to the North have included rice, cold-rolled coil, cigarette filters, chemicals, and vacuum packing equipment.

In spite of this increased exchange of goods between North and South Korea, the volume is still minimal compared with the total world trade contribution of both countries by the structure of export and import items. (See Tables 7.3 and 7.4 for the structure of export and import items of the NEACs.) Foremost among many factors currently limiting direct and indirect inter-Korean economic relations is the issue of South Korea's proposed inspections of North Korea's nuclear facilities, which would be simultaneous with, and would supplement, inspections of these facilities by the International Atomic Energy Agency (IAEA).[16]

North and South Korea will need to collaborate in order to capitalize on their various bilateral and multilateral relationships for the development and growth of their economies. South Korea, for example, could benefit by utilizing the relationships established

TABLE 7.3 Items exported to the world by north-east Asian countries, 1989 (US$ million)[a]

SITC[b]	USSR	China	Japan	South Korea	North Korea	Total
0	1 906	6 144	1 509	2 211	173	14 531
	(1.7)	(11.7)	(0.5)	(3.5)	(8.9)	(2.9)
1	288	313	140	113	1	1 202
	(0.3)	(0.6)	(0.1)	(0.2)	(0.1)	(0.2)
2	8 959	4 211	1 959	901	166	18 038
	(8.2)	(8.0)	(0.7)	(1.4)	(8.5)	(3.6)
3	6 125	4 269	955	686	153	63 385
	(56.0)	(8.1)	(0.3)	(1.1)	(7.9)	(12.7)
4	48	86	80	1	—	1 052
	(0.0)	(0.2)	(0.0)	(0.0)	(0.0)	(0.0)
5	449	3 201	14 691	204	52	25 366
	(4.1)	(6.1)	(5.3)	(3.3)	(2.7)	(5.3)
6	9 165	10 896	35 361	13 723	731	74 661
	(8.4)	(20.8)	(12.9)	(22.0)	(37.5)	(15.4)
7	22 386	3 873	19 366	23 581	151	232 502
	(20.5)	(7.4)	(70.4)	(37.8)	(7.8)	(46.4)
8	790	1 075	22 868	18 947	513	54 818
	(0.7)	(20.5)	(8.3)	(30.4)	(26.3)	(11.0)
9	9	8 733	3 804	93	6	14 531
	(0.0)	(16.6)	(1.4)	(0.2)	(0.4)	(2.5)
0–9	109 300	52 485	27 053	6 230	1 950	501 082
	(100)	(100)	(100)	(100)	(100)	(100)

[a] Figures in parentheses are percentages.
[b] SITC classifications: 0 = Food and animals; 1 = Beverages and tobacco; 2 = Raw materials; 3 = Mineral fuels and lubricants; 4 = Animal and vegetable oils and fats; 5 = Chemicals; 6 = Manufactured goods; 7 = Machinery and transport equipment; 8 = Miscellaneous manufactured articles; 9 = Miscellaneous transactions.

Source: Korea Institute for International Economic Policy (Seoul).

by North Korea with a number of countries, in terms of new markets and sources of raw materials. North Korea, for its part, could expand its access to external capital, consumer goods, and advanced technology by tapping into South Korea's foreign relationships. Furthermore, South Korean money and technological expertise could go some way towards improving North Korea's

TABLE 7.4 Items imported from the world by north-east Asian countries, 1989 (US$ million[a])

SITC[b]	USSR	China	Japan	South Korea	North Korea	Total
0	1 227	4 192	28 078	3 085	316	45 878
	(10.7)	(7.1)	(13.5)	(5.0)	(11.1)	(10.3)
1	1 896	201	2 882	186	18	4 899
	(1.7)	(0.3)	(1.4)	(0.3)	(0.7)	(1.0)
2	1 851	4 835	31 716	8 726	229	44 988
	(1.6)	(8.2)	(15.3)	(14.2)	(8.0)	(10.0)
3	5 008	1 650	43 844	7 626	916	54 342
	(4.4)	(2.8)	(15.3)	(12.4)	(32.1)	(12.2)
4	800	875	364	170	39	1 781
	(0.7)	(1.5)	(0.2)	(12.4)	(1.4)	(0.4)
5	7 926	7 556	15 070	7 153	181	37 861
	(6.9)	(12.8)	(7.3)	(11.7)	(1.4)	(8.5)
6	14 331	12 335	29 396	9 665	454	73 940
	(12.5)	(20.9)	(14.2)	(15.7)	(16.0)	(16.6)
7	58 089	1 820	28 106	21 097	561	126 055
	(50.6)	(3.1)	(13.6)	(34.4)	(19.7)	(28.3)
8	12 364	2 072	25 705	3 549	104	4 632
	(10.8)	(3.5)	(12.4)	(5.8)	(3.7)	(10.4)
9	150	7 215	2 191	117	27	9 353
	(0.1)	(12.2)	(1.1)	(0.2)	(1.0)	(2.1)
0–9	114 700	59 141	207 356	61 378	2 850	445 426
	(100)	(100)	(100)	(100)	(100)	(100)

[a] Figures in parentheses are percentages.
[b] See note *b* to Table 7.3.

Source: as for Table 7.3.

inadequate economic infrastructure, including its power distribution system and transport and communication systems, thus benefiting both parties. In fact, inadequate economic infrastructure is widely presumed to be a principal agent in North Korea's current economic difficulty, with the manufacturing sector estimated to be operating at less than 60% capacity owing to energy shortages and transportation bottlenecks.

Lastly, it can be pointed out that the new trend of internation-alization (alternatively, regionalization) of the world economy in the post-Cold War era suggests the necessity, whatever the barriers may be, for the two Koreas to find their common *raison d'être* for greater co-operation in the international arena. Ideological and political differences are giving way to economic, geographical, and national factors. It is imperative that both Korean leaderships recognize the potential of this new emerging environment for national and regional co-operation.

7.6   DEFENCE AND REGIONAL SECURITY IMPLICATIONS

Economic interaction on the peninsula will have some impact on the region's defence and security outlook. The international en-vironment and the domestic situations of the two Koreas are fac-tors that could influence short- and long-term changes in both Pyongyang and Seoul, as already discussed earlier in this chapter. A surprisingly conciliatory posture towards the United States and South Korea is reflected in the now often professed North Korean readiness to agree to mutual nuclear checks and arms control concessions in return for the US opening up of military bases in the South, reduced defence spending, the establishment of eco-nomic and diplomatic relations with Japan and the United States, and a confederal form of reunification that would leave the North's political and economic system intact.

South Korean policy-makers are beginning cautiously to examine grounds for discussion with the North on arms control. With Russian military aid no longer available and China's capabilities falling far short of military technology's cutting edge, in addition to North Korea's foreign exchange constraint, North Korea has to face a policy choice between the development of its nuclear cap-ability[17] and the proposal for unconditional simultaneous arms reductions by North and South.[18] In fact, it is generally believed that North Korea employs quite an advanced technology in weap-ons and guidance equipment and in its heavy machinery, although its technology in a number of other industries such as transport, telecommunications, and electronics is outdated.

Ironically, in their competition with one another, the two rival states have both invested too many resources in the arms race.

TABLE 7.5 Comparison of military power: North and South Korea

	North Korea	South Korea	Superiority of strength[a]
Total armed forces ('000)	1006	655	+351
Military divisions	114	65	+49
(Reserve army divisions)	(23)	(26)	(−3)
Tanks	6100	3150	+2950
(Armoured cars)	(2500)	(1600)	(+900)
Artillery	9500	4300	+5200
(Automatic weapons)	(3600)	(550)	(+3050)
Warships	460	170	+290
(Submarines)	(24)	(0)	(+24)
Tactical aircraft	850	520	+330
Support aircraft	770	690	+80
Missiles	370	270	+100
(Surface-to-surface missiles)	(39)	(19)	(+20)

[a] Superiority of North Korean over South Korean strength.

Source: Korea Institute for Defense Analyses (Seoul), Military Briefing Data, 16 July 1992.

(See Table 7.5 for the comparison of North and South Korea's military power.) The concern—indeed, fear—of a renewed Korean war has dominated their security horizon for nearly four decades. North Korea has been under threat of the South's military strength, which receives full US commitment with the nuclear umbrella. Conversely, South Korea has aimed to develop internal and external sources of strength to offset North Korea's greater military capability and to deter renewed North Korean aggression. Objectively, the record of North Korea's past behaviour has scarcely encouraged flexibility in South Korea. North Korea has sown the seeds of mistrust in the minds of Southern leaderships from President Park Chung-Hee to President Chun Doo-Hwan by attempted assassination attempts in Seoul and Rangoon, among others.

But today, few can reasonably doubt that Seoul would really be in danger from North Korea when the South possesses a far larger and more dynamic economy; today, the sheer scale of North

Korea's military capability outnumbers that of South Korea in *quantity*, but not in quality. Given North Korean across-the-board superiority in armaments, the South Korean military sector has been using a significant proportion of its national budget for some time (as already reviewed in Chapter 4) in an attempt to catch up. Seoul's security policy, and US forces in South Korea, provide military stability both in the Korean peninsula and in the rest of the region. South Korea and the United States have separately or together made a deliberate effort not to destabilize the military situation in the region. North Korea is well aware of this fact, which is why Pyongyang was led suddenly to announce, in early July 1992, that the US forces in Korea may continue to stay until the two Koreas are reunified.[19]

Thus, the existing military confrontation is likely to be gradually replaced by a movement towards an economic relationship and by interest in confidence-building measures.[20] By doing so, the two Koreas can put an end to the arms race—which is important in itself, because this should release funds and resources for economic development projects. All that is needed is a measure of mutual confidence and trust on the part of each side to take a more accommodating attitude towards the other.

7.7   KOREA TO THE YEAR 2000: THE COSTS OF REUNIFICATION

Do not be deceived by the Soviets
Do not count on the Americans
The Japanese will soon rise again
So Koreans, be careful!
Popular Korean saying after liberation

The world has grown used to the idea of two Koreas, pitted against one another in ideology and practice. But the two Korean states are now members of such major international organizations as the United Nations and its agencies. They have continued talks with one another on a regular basis following the basic agreement made on 13 December 1991. Although any tangible results are still awaited, the somewhat complementary neighbourly relations seem often to take place in a high-level political dimension. I myself am not in a position to say whether the two political leaders are making dextrous use of the continuing state of division and the stop–go talks as a means of avoiding critical domestic political and economic crises. But there is reason to wonder why it was suddenly announced that North Korea's deputy prime minister, Kim Dal-Hyun, was to visit Seoul on 19–25 July 1992, when Seoul was positively buzzing with rumours of the multi-billion *won* scandal, involving a sales contract of the land occupied by the Military Intelligence Command in Seoul. It was concurrently reported on the front pages of all Seoul newspapers that Kim's surprise visit to the South would facilitate the Daewoo participation in North Korea's Nampo port development project. (Pyongyang had agreed to set up the industrial complex in Nampo in January 1992 when the Seoul-based Daewoo Group chairman Kim Woo-Choong visited the North. But the proposed project, as well as much other inter-Korean economic co-operation, has so far faced a deadlock because of Seoul's insistence on the inspections of DPRK nuclear facilities.)

No matter what hidden politics are involved between the two rival governments, it is the genuine hope of people in both Koreas that a breakthrough will be made in achieving inter-Korean economic co-operation. Recently, the ideas for the construction of a trans-Siberian gas pipeline on the Korean peninsula and a North–South railway and expressway are being studied and supported, at least verbally, by both sides. North Korean deputy prime minister Kim Dal-Hyon's first visit to Seoul was widely reported in Seoul as bringing far-reaching changes in overall inter-Korean relations, especially in the economic and trade sectors.[21] To be sure, his trip had the result of, once again, encouraging Southern entrepreneurs to hope for opportunities to visit the North to promote economic and trade exchange, even though there remain a number of obstacles to their investment or business with North Korea, including not only the absence of a tool for direct trade and a mutual investment guarantee pact between the two Koreas, but also the North's foreign exchange shortages which are likely to hinder its settlement of trade accounts. Because of such obstacles, South Korean enterprises cannot make any positive moves with the North unless the Seoul government backs them selectively and strongly behind the scenes. It is for this reason, ironically, that North Korea is in need of working tactically with the Southern political leaders. Of course, this does not mean that politics must necessarily be independent of economics. Who can deny that a good political relationship may create a better foundation for economic co-operation, which in turn will lead to greater political harmony and perhaps even to reunification?

At the time of writing, there are still no joint ventures or pro-duction-on-order contracts between North and South Korea. Active trade relations are growing, as discussed earlier. Such trade is advantageous for both sides. So are humanitarian efforts in intra-Korean relations. The reunification of separated families is re-ferred to in recent exchanges of letters in connection with the Basic Agreement. Both governments issued their respective pro-posals in this regard, albeit quite restrictive and imprecise. It has been suggested that the inmates of ROK prisons—both ROK citizens who have been sentenced for political reasons and DPRK citizens who have been imprisoned for such things as aiding and abetting attempted espionage—be set free to choose between the North and the South. Indeed, numerous meetings are currently taking place, although with no substantial results as yet, at the official level between politicians, Red Cross members, and busi-nessmen. Meetings of this kind contribute to a reciprocal under-standing between North and South Korea, as they did between East and West Germany, despite the fact that the two states continue to differ greatly in their political views. Talks, negotiations, and agreements on tariff-free inter-Korean trade were considered inconceivable until the early 1990s; today they have become cus-tomary, generally recognized, and welcome.

However, politics and policies are always in flux. Where this movement in relations between the two Koreas will lead, no one is able to say. But it is fairly obvious that it will not lead to reunification, at least for the time being, unless some drastic 'Big Bang' occurs in the North following the death of the aged Kim Il-Sung. (It is interesting to note that the South Korean formula for reunification is a gradualist one, whereas paradoxically the North Korean formula remains abrupt and instantaneous in principle.)

7.7.1   Two Koreas to the Year 2000

The Korean economy must be analysed in relation to its reunifica-tion prospects. No one can predict when and how the two countries will be reunited and how much it will cost. Many scenarios have been suggested so far regarding reunification.[22] It could come about in a gradual process, throughout which the two governments co-operate to restore the North Korean economy to a better state. Alternatively, it may occur in a 'Big Bang' if the DPRK leadership collapses overnight in a way similar to the sensational topplings of both the Romanian leader Ceausescu and the East German leader Honecker. However, the East European-style 'people power' seems to have little chance of success in the DPRK context in the short to medium term. In any case, the possibility of such a collapse would bring about greater turmoil to the reunified economy than the German experience, because South Korea cannot be compared to the economic powerhouse of the former West Germany.

In this section, two Korean economies will be projected to the year 2000 assuming that the two states remain independent until then. Of course, the economies can change over time, in both practice and policies, so as to make such projections very vulnerable.

If North Korea does not undertake a degree of economic reform and opening up, its GNP growth rates are not expected to exceed 2% per year for the first half of the 1990s. In the late 1990s, however, the DPRK economy is expected to grow, with increasing joint ventures, at the rate of more than 6% per year if its economic co-operation with other countries improves. In terms of industrial structure, the agricultural sector's share of GNP in North Korea is expected to decline to 23% in 2000, as its economy gradually advances into a higher stage of development. Meanwhile, the service sector may expand as a result of increasing investment in infrastructure such as power generation and communications. The share of the manufacturing sector (including construction) of GNP is projected to remain steady. Per capita GNP, using the trade exchange rate (Won 2.15 to the US dollar) as a conversion factor, is projected to be about $1.413 in 2000.

In passing, it should be noted that North Korea undertook a sudden reform of its banknotes on 15 July 1992. A government decree carried by the official Korean Central News Agency (KCNA) said that the measure was aimed at strengthening the independent monetary system of the country and further facilitating monetary circulation. The existing W100, W50, W10, W5, and W1 notes were replaced by new notes of the same denominations.[23] It is believed that the measure was carried out to draw out private savings of cash that had effectively been withdrawn from the money

TABLE 7.6 Projection of major economic indicators of North Korea, 1990–2000 (in 1990 prices)

	1990	1995	2000
Population (million)	22.9	24.9	26.9
(Growth rate) (%)	(1.8)	(1.7)	(1.6)
Workforce (million)	10.3	11.9	13.1
Labour force participation (%)	72.0	75.0	78.0
GNP (US$ billion)	25.7	28.4	38.0
(GNP growth rate) (%)	(–3.7)	(2.0)	(6.0)
Industry structure (% of GNP)			
Agriculture	26.8	25.0	23.0
Mining and manufacturing	56.0	56.0	54.0
Services	17.2	19.0	23.0
GNP per capita[a] (US$)	1122	1141	1413
Exports (US$ billion)	2.02	2.4	4.0
(% of GNP)	(7.9)	(8.6)	(10.6)
Imports (US$ billion)	2.62	2.7	4.2
(% of GNP)	(10.2)	(9.4)	(11.0)

[a] Based on trade exchange rate, projected to be US$1 = W2.15.

Sources: Tables 3.5 and 3.6; also, Korea Development Institute; South Korean National Unification Board.

supply. The previous notes had been legal tender since 1979. Issuing the new currency was indeed aimed at boosting government liquidity and ensuring a better control of the money supply, thereby enabling better management of the overall economy.

This measure constitutes one of the economic options available to the North Korean leadership to forcibly mobilize domestic (private) savings for national investment. North Korea is badly in need of investment in its inadequate economic infrastructure as well as the development of the light industry sector with whatever sources of available capital it can mobilize in the short and medium terms.

With regard to trading patterns, China would continue to be North Korea's largest trade partner under current conditions, with Japan becoming its most important market during the 1990s. If the CIS economies recover, then DPRK trade with them may increase, as it may with South Korea as the DPRK–ROK relationship improves.

For South Korea, the major challenges appear to lie in pursuing economic restructuring and industrial relocation subject to changing international economic environments. The ROK has kept to a path of rapid economic growth in the past, but at present the country is, somewhat ironically, sandwiched between the advanced countries and just-behind countries, such as China, Thailand, and Malaysia. Domestic politics has also eroded economic vitality and entrepreneurial volition in the management of the macroeconomy during the period of the Sixth Republic (1986–92). The continued pressures for democratization, and the dramatic social transformation resulting from the rapid promotion of self-interest, have soaked up the society's work ethics and morality, which had played a very substantial role in ROK economic development for four decades. Productivity has grown relatively less than wage increases in all productivity and service sectors. To make matters worse, favourable international environments have altered, thus bringing the South to a crossroads in its economic and social progress. South Korean economic development has so far been heavily dependent upon the rapid growth of international trade, being particularly reliant on exports. The country enjoyed large surpluses in its current account for three years after 1985 because of lower oil prices, the appreciation of the Japanese yen, and a decline in world interest rates. Nevertheless, the economy failed to take advantage of this opportunity; instead, South Koreans pursued an idle and luxurious lifestyle. Thus, the balance of trade has been suffering from decreasing competitiveness, owing to an increase in the demand for imported luxury goods and also a rapid increase in domestic wage levels.

Subject to external economic conditions, the South Korean economy needs to undertake fundamental restructuring as well as industrial relocation. In the search for a lower-wage work-force and cheaper industrial sites, the relocation process will focus on China, Indochina, south-east Asia, and North Korea. Seoul will also have to upgrade its industrial base in order to develop a comparative advantage in capital- and knowledge-intensive industries. The current transition may necessarily slow the rate of economic growth in the early 1990s; nevertheless, the country's seventh Five-Year Economic and Social Development Plan (1992–6) proposes

TABLE 7.7 Projection of major economic indicators of South Korea, 1990–2000 (in 1990 prices)

	1990	1995[a]	2000
Population (million)[b]	43.5	45.4	47.4
Employment (million)	18.0	20.4	22.2
Unemployment rate (%)[c]	2.4	2.4	2.4
GNP (US$ billion)[d]	245	493	731
Industry structure (% of GNP)			
Agriculture	18.3	14.0	10.0
Manufacturing	27.3	26.0	25.0
SOC and others	54.4	60.0	65.0
Per capita GNP (US$ '000)	5.7	10.9	15.4
Exports (US$ billion)	65	120	193
Imports (US$ billion)	70	117	188

[a] Figures for 1995 are obtained using the rates used in the Plan.
[b] Population growth rate is assumed to be 0.85% per year from 1996 to 2000.
[c] Employment is assumed to be 44% and 47% of total population in 1995 and 2000 respectively.
[d] The growth rate of GNP is assumed to be 6% a year after 1996 to 2000.

Sources: Economic Planning Board, *The Seventh Five-Year Economic and Social Development Plan 1991*; Bank of Korea, *Economic Statistics Yearbook* (1991).

a growth rate of 7.5% per year. The rate is expected to drop slightly to about 6% annually after that to the year 2000. Assuming that there is no reunification until at least 2000, the economy is projected as presented in Table 7.7.

7.7.2   The Cost of Reunification

The cost of reunification may vary according to the time, the patterns, and the circumstances under which it is achieved. Undoubtedly, it will not be inexpensive. The cost must include not only the investment required to restructure industries as a result of reunification, but also the internal and external costs involved in socio-economic integration and transition.

With regard to timing, two main streams of opinion prevail. Officially, the ROK government prefers a gradual process of controlled or managed reunification, with the major emphasis on the development of economic links on the German model. Neither does the DPRK leadership want to see the East German experience repeated on the Korean peninsula, even though its leadership has maintained that the South Korean regime must be overthrown overnight. If we reasonably eliminate any possibility of reunification on the North's terms, the cost will be largely related to the timing, the extent of socio-economic friction, and the ease and the speed of integration.

Neoclassical economists tend to assume that human adjustments to new situations will be essentially swift and basically unproblematic. The possibility of slow and costly adjustments to changes—or high friction—while acknowledged, is treated as rare. From this perspective, neoclassicists have advocated 'shock therapy' for post-communist countries—that they jump straight into a free market from a command-and-control system.[24]

Seen from this view, this would be the best, or the least-cost, way for the two Koreas to bring about reunification as quickly as possible, without taking account of the existing chasm between the two economies. On the other hand, the 'gradualist' proponents define the cost of reunification as the amount of investment that would be required either to restructure North Korean industries or to equalize per capita GNP of the North with that of South Korea. A recent study to put a figure on it by the Korea Development Institute suggested a cost of $140 billion for the first three years alone and a total of $250–$300 billion by the end of the decade. The cost estimates vary because of the differential assumptions regarding the prospects of North Korea's economic reform.

Alternatively, the amount of investment required to equalize the per capita GNP of North Korea with that of the South can be calculated by the marginal capital–output ratio for North Korea and the per capita income difference between the two countries. The problem is how to obtain the data to estimate the marginal capital–output ratio for the North Korean economy.

The estimated values of the capital–output ratio for many income cohorts' countries range between 2.5 and 3.5 depending on the levels of capital intensity for the economies under study. South Korean economic data show that this ratio has been quite stable at around 3.0–3.2 for the past 30 years. Here the same marginal capital–output ratio is assumed applicable to the North Korean economy.[25]

In 1990 the per capita GNP of South Korea was $5659, and that of North Korea was estimated to be $1122, using the trade exchange rate as a conversion factor. The difference is $4537. Considering that North Korea's population was 22.9 million, the required rise in GNP to equalize the per capita GNPs of the two Koreas is approximately $103.9 billion. Using the capital–output ratio of 3.0 (or 3.2), the required increase in capital stock or investment is $312 (or $332) billion. In other words, the cost of unification would be approximately $312–$332 billion, which is about 1.3 times the GNP of South Korea in 1990.

In 1995, the per capita GNP of South Korea is projected to be $10 900 while that of North Korea is expected to be $1141. The difference is $9759. With a projected population of 24.9 million, the required increase in North Korea's GNP to equalize the per capita GNPs of the two countries would be about $243 billion. Applying the capital–output ratio of 3.2, the required increase in capital stock is $777.6 billion. In the year 2000, GNP per capita is projected at $15 400 for South Korea as contrasted to $1413 for the North. The difference amounts to $13 987. With North Korea's population projected at 26.9 million, the required increase in GNP to equalize per capita GNPs on the peninsula reaches approximately $376.3 billion. Using the marginal capital–output ratio of 3.2 factor, the required incremental investment is about $1204.0 billion, which is about 1.6 times the expected GNP of South Korea in 2000.[26]

In sum, the cost of reunification depends largely upon the expected per capita income gap between the two Koreas, the total population of the country that has the relatively lower per capita income, and the factor of marginal capital–output ratio of that country. But it must be noted that this kind of cost estimate tends to underestimate the real cost because of not taking into account the cost incurred in psychological, sociological, and political factors. Assuming the adjustment costs for socio-economic externality to be about the same as the investment cost required for a per capita income equalization programme, the total reunification cost would be approximately $2408.0 billion in the year 2000.[27] The larger the income gap between the two states and the greater the friction factors, the greater the costs of reunification will be.

In concluding this section, it must be noted that there are several socio-economic factors that suggest that reunification will have to wait until North Korea can increase its growth rate as well as opening up towards a more democratic and market-oriented way of living. Such a transformation would include human factors, capital, and enterprises under non-privatized state control, infrastructure, labour mobility and social values. The discussion of such issues is beyond the scope of this book; but it is worth mentioning, among other things, that the transition to a capitalist style of work and competition will be very costly and difficult for those who have acquired specific personality traits and work habits that cannot be modified in short order. In addition, large amounts of the capital and many state enterprises as well as military assets accumulated in communist economies are neither transferable nor saleable in market-oriented economies. These are some examples of the factors to be taken into consideration when attempting to estimate the cost of reunification.

It is clear that the costs of transition and of reunification are going to be much higher than at first thought. What is now required is a lowering of expectations and an acceptance of the need for a long period of gradual transition leading up to an eventual reunification of North and South Korea.

NOTES TO CHAPTER 7

1. Jane J. Mansbridge (ed.), *Beyond Self-Interest* (Chicago: University of Chicago Press, 1990), 97.
2. Su Tung-Pao (1036–1101), *Jun-Jokbyokboo* (Pre-Red Wall Thought).
3. Yu Chin-O, 'Turning Period and Spiritual Direction', in *Direction of New Generation* (in Korean) (Seoul: Hak-Won-Sa, 1964).
4. William Nordhaus, 'The Longest Road: From Hegel to Haggle', paper presented at the Brookings Panel on Economic Activity, Washington, DC, 5–6 April 1990.
5. On 7 July 1992, South Korea proposed that the North and South allow dispersed families to meet on a regular basis and choose to settle on either side according to their 'free will'. The proposal was made by South Korea's Prime Minister Chung Won-Shik in a letter sent to his northern counterpart Yon Hyong-Muk through the truce village of Panmunjom at 10.00 am on 7 July 1992.
6. Lee In-Mo is known to have been a military journalist during the Korean War and engaged in partisan activities against South Korea even after the war. He served a long prison term and was released without a change in his ideological allegiance. He had been living with a farm family in Masan City, South Korea, while his wife and daughter's family members are living in North Korea. In his 7 July message to North Korea, the South's Prime Minister Chung suggested that the South would be willing to allow this 75-year-old defiant communist, Lee In-Mo, to return to the North if the North would return those who were abducted by the North against their own will. But in March 1993, newly elected President Kim Young-Sam of South Korea sent Lee In-Mo to North Korea in a humanitarian gesture.

7. Since the two Koreas first initiated contact with each other via Red Cross talks in 1971, political, economic, and humanitarian contact has been maintained in a variety of official, unofficial, open and secret ways. However, the achievements of these contacts after more than 22 years are exceedingly meagre.

8. The North Korean Workers' Party, through its newspaper, *Rodong Sinmun*, on 21 September 1991 warned the people of the dangers of the capitalist market system, saying: 'The criticism of the planned socialist economy by the imperialists is based on a false sophistry designed to destroy the socialist system and thus to place it under their subordination. The capitalist market system, however it may be beautified, is nothing but an anti-people economic system which is aimed at exploiting and oppressing the masses and to bring about unemployment and poverty.'

9. One could characterize change in communist regimes as follows:

   • The China model (*perestroika* without *glasnost*)
   • The USSR model (*glasnost* without *perestroika*, which brings about the collapse of the system)
   • The Polish/Czech model (systematic collapse of the system, as a result of structural friction when market factors are introduced in the command economy)
   • East German model (systematic collapse with external capitalist takeover).

10. These economic zones include Zhenzhen, Zhuhai, Shantou, Xiamen, and Hainan. The main aim of setting up these zones is to utilize their geographical proximity to Hong Kong, Macao, and Taiwan, to absorb more capital and advanced technologies from overseas, to develop foreign trade, and to promote economic construction. Special policies are carried out in these zones which are not condoned by the existing system of the country, in the form of ownership of the means of production, and the management and administration of the enterprises.

11. Tim Dunk, 'Prospects For Inter-Korea Cooperation', paper prepared by the National Korean Studies Centre as an input to the seminar, 'Korea to the Year 2000: Implications for Australia and Policy Responses', carried out by the East Asia Analytical Unit of the Australian Department of Foreign Affairs and Trade.

12. Japanese Prime Minister Kiichi Miyazawa said, on 2 July 1992, that the USA, Japan, China, and Russia should co-operate to support dialogue between South and North Korea. Easing tension on the Korean peninsula was the most crucial task for the security of Asia and the Pacific, Miyazawa told the National Press Club in Washington, DC.

13. North-east Asian economic co-operation has been an important topic among many economists and politicians in recent years. Particularly with the imminent birth of the European Integration and the North American Free Trade Agreement (NAFTA), the topic became more popular among north-east Asian countries which encompass the states of Japan, South Korea, North Korea, Mongolia, China, and the Far Eastern portion of the former USSR.

14. M. Miller, A. Holm, and T. Kollohor, 'Tumen River Area Development Mission Report', report of the UN Development Programme presented in Pyongyang (16–18 October 1991), 2–4.

15. The Korean question is not just about North and South. China, Japan, the former USSR, and the USA have affected the life of Koreans in their modern history. Some of these countries are still crucial intermediaries between the two governments on the peninsula.

16. The Japanese government also made it clear that there could not be a normalization of relations between Tokyo and Pyongyang as long as international

suspicion concerning North Korea's nuclear programme was unresolved. Japanese Prime Minister Kiichi Miyazawa assured US President George Bush of this at their meeting at the White House on 2 July 1992. It is remarkable that the USA is more bent on pressuring North Korea into mutual inspections than South Korea is.

17. The information so far available suggests that Pyongyang found it much more difficult and much more expensive than expected to pursue its drive for a nuclear weapons option without foreign financial and technical help. Thus, it finally decided to suspend the effort at the Workers' Party Central Committee meeting in December 1991.

18. This would seem to have been part of a larger policy controversy between insular old elements and more outward-looking elements in the North's leadership who want to resolve domestic economic problems by promoting foreign economic co-operation and a reduction of defence spending linked to arms and forces reduction agreements with the South.

19. North Korea has recently removed an article on communizing South Korea from its constitution, a senior fellow at the Carnegie Endowment for International Peace who visited North Korea 28 April–4 May 1992 told a seminar given at the Hilton Hotel in Seoul on 3 July 1992. Quoting a high-ranking North Korean official whom he wasn't free to identify, Harrison said that the Workers' Party in the North recently modified its long-standing position that asserted to drive foreign forces out of the country on a 'nationwide scale'. On 6 July 1992, the major Japanese newspaper *Mainichi* quoted a senior US official as saying that North Korea would allow US troops in South Korea to be gradually withdrawn after the two Koreas were reunited. The official said that the intentions were conveyed by North Korean Workers' Party Secretary for International Affairs Kim Yong-Sun when he held unprecedented talks in January 1992 with US Under Secretary of State for Political Affairs Arnold Kanter in New York.

20. Given the current opportunity for working together on the Tumen River project and other regional economic co-operation, both North and South Korea are in a good position to pursue common interests that will help erect a framework for confidence-building measures.

21. The chronology of the Seoul initiative for ROK–DPRK Economic Exchanges is as follows:

7 July 1988	South President Roh Tae-Woo announced a special declaration for direct inter-Korean economic exchanges and co-operation.
18 October 1988	The South announced guidelines on South–North Korean trade underlining the permission of private trade between South and North Korean businesses.
January 1989	South Korea's Hyundai Group founder Chung Ju-Yung visited Pyongyang and agreed with North Korea's President Kim Il-Sung to develop Mt Kumgang in the North. The joint development project was soon aborted owing to strained relations between Seoul and Pyongyang following illegal trips to the North by the South's student activist coed Im Su-Kyong and Revd Moon Ik-Hwan.
January 1990	Seoul's Economic Planning Board sets up a fund for inter-Korean economic co-operation.
13 April 1991	Formal direct trade is realized between the South's rice and the North's cement and anthracite coal.
17 September 1991	Entry of both North and South Koreas into the United Nations.

October 1991	A six-nation conference involving South and North Korea on the development of the Tumen River estuary as a free trade and industrial zone was held in Pyongyang, initiated by the UN Development Programme.
13 December 1991	Pyongyang and Seoul signed an agreement concerning inter-Korean exchange and co-operation.
15 January 1992	Daewoo Group chairman Kim Woo-Choong visited Pyongyang and met Kim Il-Sung.
19–25 July 1992	North Korean Deputy Prime Minister Kim Dal-Hyon, accompanied by 9 economic technocrats, visited Seoul and met Mr Roh Tae-Woo, who was to leave his presidency in less than seven months.

22. Tim Dunk, 'The Two Koreas: Total Reunification?' paper presented at the seminar, 'Korea to the Year 2000', Seoul National University, 7 July 1992.
23. It is known that Pyongyang's central bank was willing to exchange old money for new at the one-for-one rate for only six days after the 15 July deadline.
24. Amitai Etzioni, *Eastern Europe: The Wealth of Lessons* (Washington, DC: George Washington University, 1991).
25. See Dunk, 'The Two Koreas', 5.
26. See also ibid.
27. Note that this reunification cost will be less than half of this figure when it is calculated by North Korea's dollar GNP, which is based on the official exchange rate instead of the trade exchange rate (see Chapter 3).

Epilogue

Of making many books there is no end,
    and much study wearies the body.
Now all has been heard;
Here is the conclusion of the matter;
Fear God and keep His commandments,
    for this is the whole duty of man.
For God will bring every deed into judgment,
    including every hidden thing,
Whether it is good or evil.

Eccles. 12: 12–14

This book was largely written in 1990 when I was a MacArthur Scholar at the Brookings Institution in Washington, DC. After the first six chapters were completed, there were many further developments within and around the Korean peninsula.

First of all, North Korea decided to apply for a separate UN seat simultaneously with South Korea. In an announcement made on 27 May 1991, the Pyongyang Foreign Ministry said that North Korea could not help but take this action in order to 'find a way out of the difficult situation of today which has been created by the South Korean authorities'. The Pyongyang leadership had claimed until then that simultaneous entry would consolidate permanently the division of the peninsula. In that sense, North Korea's application signalled a big swing in its foreign policy. The UN Security Council decided to accept the two Koreas as members on 9 August 1991, and they became UN members on 17 September 1991.

On the domestic front, there have been many new inter-Korean developments. One important event was that an initial shipment of 5000 tons of South Korean rice was delivered to the North in July 1991, marking the opening of direct trade between the two countries. In return, North Korea shipped 30 000 tons of anthracite coal and 11 000 tons of cement to the South. In addition, there have been several other exchanges and talks between the two

Koreas, eventually leading to the Basic Agreement on Peace, Reconciliation, and Trade between the two states.

Although there has been an increase in the trend of trade with one another, substantial relations appear to have stalled because of North Korea's nuclear weapons programme. How the Pyongyang leadership will respond to world pressure to give up its nuclear arms stockpiles is yet to be seen. But it is likely to open its nuclear arms facilities for inspection, as it is badly in need of economic co-operation from South Korea, Japan, and the United States.

Indeed, the political and economic climate enveloping the Korean peninsula and the world as a whole is changing so rapidly that any rational forecast of the future is literally impossible, and therefore any manmade model or scenario about future events is increasingly unrealistic. It can only be hoped that inter-Korean relations develop to a greater degree and in a better way in all aspects of inter-action. All Koreans would like to see such relations lead to eventual reunification regardless of the cost. But who on earth can ignore the Proverbs saying that 'in his heart a man plans his course, but the Lord determines his steps' (Prov. 16: 9)?

In concluding, I must stress that this book cannot claim to have answered every important question on the Korean economies. Nevertheless, I hope that it will provide extensive material for comparative studies of the two Koreas. I myself feel some in-adequacy in covering such an important and topical subject, but I promise to continue to work on it.

APPENDIX 1

Agreement on the Implementation of Trade and Economic Co-operation and the Establishment of the South–North Joint Economic Co-operation Committee between the South and the North (Draft)

The two sides,

Desiring to implement direct material trade and programmes of economic co-operation on the basis of reciprocity and equality,

Recognizing that the implementation of direct trade and economic co-operation will contribute to the prosperity and welfare of our peoples, and the peaceful reunification of our nation;

Being assured that trade and economic co-operation shall be implemented in good faith and with sincerity,

Have agreed as follows on the implementation of trade and economic co-operation and the establishment of the South–North Joint Economic Co-operation Committee:

1. The two sides shall take all necessary and appropriate measures to implement and expand trade and economic co-operation effectively between them.
2. The two sides shall designate, for each commodity and project, appropriate trade organizations, corporations, associations, or other authorities to participate in commodity trade and economic co-operation projects.
3. Trade between the two sides shall be implemented according to the following guidelines except in cases where the two sides agree on a different procedure:
    (a) The two sides shall start trade with the commodities specified below and may expand trade to other commodities by common consent.
        (i) Commodities to be sold by the South: iron products, textiles, salt, tangerines, and such South Sea marine products as green seaweed, brown seaweed, oysters, and anchovies.

(ii)   Commodities to be sold by the North: anthracite coal, iron ore, magnesia clinker, pollack, and corn.

(*b*)   The amount and volume of trade shall be decided through consultations of the designated parties concerned, considering the supply and demand of the commodity.

(*c*)   The price of the commodity shall be decided by the designated parties concerned, considering international market prices.

(*d*)   The two sides shall conduct trade by means of back-to-back letters of credit issued from third-country banks. However, in cases where the amount and value of commodities to be traded is the same, the two sides may conduct trade by means of barter trade without the exchange of letters of credit or notes.

(*e*)   A third-country bank, to be agreed upon by the two sides, shall settle the accounts for the letters of credit.

(*f*)   The currencies for settlement shall be the British pound and the Swiss franc.

(*g*)   The two sides shall not impose customs or similar charges which they impose on imports from other countries, on the commodities they purchase from the other side.

(*h*)   As for customs procedures, inspections, dispute settlements, etc., in connection with commodity trade between the two sides, they shall apply the same regulations which they use for normal external trade.

4. The two sides shall reconnect the Seoul–Shinuiju railway line to facilitate the smooth transport of commodities for trade and economic co-operation.

5. The mode of transportation for commodities shall be determined through consultations between the designated parties, considering the character, weight, etc., of the commodities involved and the costs of transport. In cases of marine transport, the two sides shall guarantee to extend the most favorable treatment to the transport ships of the other side regarding entry, anchoring, unloading and loading, departure, etc., and to handle all procedures promptly.

6. The two sides shall implement joint economic projects to promote the common prosperity of the peoples of both sides, and shall start joint economic projects in the areas specified below and expand projects into other areas by common consent.

(*a*)   The establishment of joint fishing areas.

(*b*)   The joint development of natural resources.

7. The scale, method, conditions, timing, etc., for the implementation of the joint projects shall be determined through consultations of the designated parties concerned.

8. The two sides may, if agreed upon, extend exemption or reduction of income taxes, corporation taxes, property taxes, customs, and other taxes for the designated parties concerned of the other side when the designated parties concerned are operating in their territories.

9. The two sides shall open the ports of Inchon and Pohang in the South, and Nampo and Wonsan in the North, in order to facilitate marine transport for commodity trade and joint economic projects. In addition, as trade and economic co-operation continue to expand, other ports may be opened as agreed upon.

10. The two sides shall establish communication facilities necessary for trade and joint economic projects, and these facilities may be expanded as agreed upon.

11. The two sides shall permit the designated parties concerned of the other side to visit their territories or territorial waters in order to conduct advance inspections for commodity trade, feasibility studies for joint projects, and related activities.

12. The two sides shall guarantee the visits and safety of the persons concerned with trade and economic co-operation of the other side and assist them as much as possible with traffic, communications, lodging and boarding, medical care, etc.

13. The two sides shall take prompt and effective relief measures for physical injuries incurred by the persons concerned of the other in one's area and notify the other side without delay of the details of the situation.

14. The two sides shall, within thirty days after the signing of the agreement, establish and operate the South–North Joint Economic Co-operation Committee (hereafter referred to as the 'Joint Committee'), chaired by deputy prime minister-level officials of the two sides. The Joint Committee shall meet in order to carry out this agreement, to discuss and decide on the methods for developing commodity trade and joint economic projects, and to ensure the implementation of all decisions.

15. The organization of the Joint Committee and supporting bodies shall be as follows:

    (a) The Joint Committee shall consist of seven members from each side: one chairman at the deputy prime minister level; one vice chairman at the ministerial level; and five other members at the ministerial or vice ministerial level selected from among the officials of the government and economic circles.

    (b) Under the Joint Committee, there shall be the Sub-Committee for Commodity Trade and the Sub-Committee for Economic Co-operation and there may be, if necessary, other special Sub-Committees through mutual agreement. To ensure the smooth operation of the Joint Committee and sub-committees, there may be small sub-committees.

(c) Each sub-committee shall consist of five persons from each side including the chairman. Sub-committee chairmen shall be appointed by respective Joint Committee chairmen from among Joint Committee members. Sub-committee members shall be at the bureau director level.

(d) Under the Joint Committee, there shall be a Joint Secretariat charged with working-level matters. The secretary-general of the Joint Secretariat shall be appointed by the chairman of the Joint Committee from among the members of the Joint Committee. At the Secretariat, there shall be clerical officials of a number agreed on by the two sides. The location of the Secretariat shall be Panmunjom.

16. The Joint Secretariat shall execute the following functions:

(a) The faithful implementation of an agreement on commodity trade and economic co-operation projects between the South and the North.

(b) Discussion and determination of measures to expand and develop commodity trade and economic co-operation projects between the South and the North. Guaranteeing of its implementation and the conclusion of necessary agreements.

(c) Discussion and co-ordination of the problems arising from the implementation of an agreement on commodity trade and economic co-operation projects between the South and North.

(d) Taking of necessary administrative measures to guarantee the safe transport of trading commodities and free travel of people concerned in the implementation of commodity trade and co-operation projects between the South and the North.

(e) Promotion and encouragement of the mutual exchanges of persons, trade fairs, shows, information, materials, etc., related to commodity trade and economic co-operation projects between the South and the North.

(f) Discussion and determination of other necessary measures to promote commodity trade and economic co-operation projects between the South and the North.

17. Sub-committees shall execute the following functions:

(a) Discussion and determination of concrete methods for the implementation of measures agreed to or mandated by the Joint Committee and ensuring their implementation.

(b) Preparation of a draft agreement necessary for the concrete realization of commodity trade and economic co-operation projects, and submitting it to the Joint Committee.

    (*c*)   Discussion and settlement of problems arising in their respective areas and presentation of them, if necessary, to the Joint Committee.

18. The Joint Secretariat shall execute the following functions:
        (*a*)   Provision of all necessary administrative assistance to the Joint Committee and sub-committees in the form of arranging committee meetings, preparing the place and agenda for meetings, and recording.
        (*b*)   Establishment and operation of a joint commodity exchange for the exchange of samples, material, and letters related to commodity trade and economic co-operation projects.
        (*c*)   Execution of liaison services and the provision of administrative support necessary for the transport of commodities and travel of people between the two sides.

19. The Joint Committee and sub-committees shall be operated as follows:
        (*a*)   The meetings of the Joint Committee and the sub-committees shall be held at Panmunjom, and may be held in Seoul and Pyongyang as agreed upon between the two sides.
        (*b*)   The regular meetings of the Joint Committee shall in principle be held every three months and the meetings of the sub-committees whenever deemed necessary as agreed upon between the two sides.
        (*c*)   The meetings of the Joint Committee and sub-committees shall be held in principle behind closed doors. However, such meetings may be opened to the public under mutual agreement if necessary.
        (*d*)   The trading parties and the parties to economic co-operation projects of the two sides and the secretary-general and clerical officials of the Joint Secretariat may be allowed to attend the meetings of the Joint Committee and sub-committees.
        (*e*)   Other matters necessary for the operation of the Joint Committee shall be determined separately through consultations.

20. The Joint Secretariat shall be established as follows:
        (*a*)   The Joint Secretariat shall be established within thirty days after the signing of this agreement.
        (*b*)   The Joint Secretariat shall use the House of Peace and Panmungak as temporary offices pending the time of the construction of the exclusive office building of the Joint Secretariat at Panmunjom.
        (*c*)   The expenses necessary for the construction and operation of the exclusive office building of the Joint Secretariat shall be borne jointly.

21. This agreement may be amended or supplemented through mutual agreement.
22. This agreement shall be valid for a period of five years, from the date of its entering into force. Upon the expiration of the said period, its validity shall be automatically extended for another period of five years unless either of the two sides notifies the other side of its intention to terminate this agreement one year prior to its expiration.
23. This agreement shall enter into force on the date when it is signed and authentic texts are exchanged.

In witness whereof, the undersigned, duly authorized by the highest authorities of their respective sides, signed this agreement and exchanged the equally authentic duplicate copies of the agreement done on June 20, 1985, at Panmunjom.

APPENDIX 2

Full Text of the Agreement on Reconciliation, Non-aggression, and Exchanges and Co-operation
(13 December 1991)

Whereas, in keeping with the yearning of the entire people for the peaceful unification of the divided land, the South and the North reaffirm the unification principles enunciated in the July 4 [1972] South–North Joint Communique;

Whereas both parties are determined to resolve political and military confrontation and achieve national reconciliation;

Whereas both desire to promote multifaceted exchanges and co-operation to advance common national interests and prosperity;

Whereas both recognize that their relations constitute a special provisional relationship geared to unification; and

Whereas both pledge to exert joint efforts to achieve peaceful unification;

Therefore, the parties hereto agree as follows:

SOUTH–NORTH RECONCILIATION

Article 1: The South and the North shall respect each other's political and social systems.

Article 2: Both parties shall not interfere in each other's internal affairs.

Article 3: Both parties shall not slander and vilify each other.

Article 4: Both parties shall not attempt in any manner to sabotage and subvert the other.

Article 5: Both parties shall endeavor together to transform the present armistice regime into a firm state of peace between the South and the North and shall abide by the present Military Armistice Agreement (July 27, 1953) until such time as such a state of peace has taken hold.

Article 6: Both parties shall cease confrontation on the international stage and shall co-operate and endeavor together to promote national interest and esteem.

Article 7: To ensure close consultations and liaison between both parties, a South–North liaison office shall be established at Panmunjom within three months of the effective date of this agreement.

Article 8: A South–North Political Sub-committee shall be established within the framework of the Inter-Korean High-Level Talks within one month of the effective date of this agreement with a view to discussing concrete measures to ensure the implementation and observance of the accords on South–North reconciliation.

SOUTH–NORTH NON-AGGRESSION

Article 9: Both parties shall not use armed force against each other and shall not make armed aggression against each other.

Article 10: Differences of opinion and disputes arising between the two parties shall be peacefully resolved through dialogue and negotiations.

Article 11: The South–North Demarcation Line and areas for non-aggression shall be identical with the Military Demarcation Line specified in the Military Armistice Agreement of July 27, 1953 and the areas that have been under the jurisdiction of each party respectively thereunder until the present.

Article 12: To abide by and guarantee non-aggression, the two parties shall create a South–North Joint Military Committee within three months of the effective date of this agreement. The said Committee shall discuss and carry out steps to build military confidence and realize arms reductions, including the mutual notification and control of major movements of military units and major military exercises, the peaceful utilization of the Demilitarized Zone, exchanges of military personnel and information, phased reductions in armaments including the elimination of weapons of mass destruction and surprise attack capabilities, and verifications thereof.

Article 13: A telephone hot-line shall be installed between the military authorities of both sides to prevent accidental armed clashes and avoid their escalation.

Article 14: A South–North Military Subcommittee shall be established within the framework of the Inter-Korean High-Level Talks within one month of the effective date of this agreement in order to discuss concrete measures to ensure the implementation and observance of the accords on non-aggression and to resolve military confrontation.

SOUTH–NORTH EXCHANGES AND CO-OPERATION

Article 15: To promote an integrated and balanced development of the national economy and the welfare of the entire people, both parties shall conduct economic exchanges and co-operation, including the joint development of resources, trade in goods as a kind of domestic commerce, and joint investment in industrial projects.

Article 16: Both parties shall carry out exchanges and co-operation in diverse fields, including sciences, technology, education, literature, the arts, health, sports, the environment, and publishing and journalism, including newspapers, radio, television, and publications in general.

Article 17: Both parties shall guarantee residents of their respective areas free inter-Korean travel and contacts.

Article 18: Both parties shall permit free correspondence, reunions, and visits between family members and other relatives dispersed South and North, shall promote the reconstitution of divided families on their own, and shall take measures to resolve other humanitarian issues.

Article 19: Both sides shall reconnect railroads and roads that have been cut off and shall open South–North land, sea, and air transport routes.

Article 20: Both parties shall establish and link facilities needed for South–North postal and telecommunications services and shall guarantee the confidentiality of inter-Korean mail and telecommunications.

Article 21: Both parties shall co-operate on the international stage in the economic, cultural, and various other fields and carry out joint business undertakings abroad.

Article 22: To implement accords on exchanges and co-operation in the economic, cultural, and various other fields, both parties shall establish joint committees for specific sectors, including a South–North Economic Exchanges and Co-operation Committee, within three months of the effective date of this agreement.

Article 23: A South–North Exchanges and Co-operation Sub-committee shall be established within the framework of the Inter-Korean High-Level Talks within one month of the effective date of this agreement with a view to discussing concrete measures to ensure the implementation and observance of the accords on South–North exchanges and cooperation.

AMENDMENTS AND EFFECTUATION

Article 24: This agreement may be amended or supplemented by concurrence between both parties.

Article 25: This agreement shall enter into force as of the day both parties exchange instruments of ratification following the completion of their respective procedures for bringing it into effect.

Date: December 13, 1991

Chung Won-shik	Yon Hyong-muk
Prime Minister	Prime Minister
Republic of Korea	Administration Council
	Democratic People's
	Republic of Korea

SOUTH–NORTH JOINT STATEMENT

[Following is a joint statement issued by Lee Dong-bok, spokesman for the South Korean delegation, and his North Korean counterpart, An Byong-su.]

1. The South and the North signed the Agreement on Reconciliation, Non-aggression, and Exchanges and Co-operation December 13, 1991, and agreed to complete as soon as possible their respective procedures necessary for putting the accord into effect.
2. The South and the North, sharing the view that no nuclear weapons should exist on the Korean Peninsula, agreed to hold a meeting by the end of December at the truce village of Panmunjom to discuss the nuclear issue.
3. The South and the North agreed to hold the sixth round of Inter-Korean High-Level Talks February 18–21 next year [13 December 1993] in Pyongyang.

Bibliography

ADELMAN, IRMA (ed.), *Practical Approaches to Development Planning: Korea's Second Five-Year Plan* (Baltimore: Johns Hopkins University Press, 1969).

AKIMO, HIDEO, 'On the Economy' and 'Topics in Heavy Industry', *Record of North Korea* (Tokyo, 1960), 75–118.

AHY KWANG-ZUP, *Socialistic Accumulation in the People's Economy*, (Pyongyang: Academy of Social Sciences, 1964).

ASMUS, RONALD D., 'A United Germany', *Foreign Affairs*, 69:2 (1990), 63–76.

BAEK JONG-CHUN, *Problem for Korean Unification* (Seoul: Research Centre for Peace and Unification, 1988).

BENASSY, JEAN-PASCAL, *Macroeconomics: An Introduction to Non-Walrasian Approach* (New York: Academic Press, 1986).

BLINDER, A. S., and R. M. SOLOW, 'Does Fiscal Policy Matter?' *Journal of Public Economics*, 2 (1973), 319–37.

BOFINGER, PETER, 'The German Monetary Unification (GMU): Converting Marks to D-Marks', *Federal Reserve Bank of St Louis Bulletin*, 72:4 (1989), 17–35.

BUNGE, FREDERICA M. (ed.), *North Korea: A Country Study* (Washington, DC: US Government Printing Office, 1981).

CASTRO, FIDEL, 'Opening Address to the Second Congress of the Association of Third World Economies', Havana, 26 April 1981.

CHENERY, HOLLIS B., and STROUT, A., 'Foreign Assistance and Economic Development', *American Economic Review*, 56 (1966), 680–733.

CHO MYUNG-KYUN, *Explanation of North Korea's GNP Estimation Approaches* (Seoul: National Unification Board, 1988).

CHOI HO-CHIN, 'The Process of Industrial Modernization in Korea: The Later Part of the Chosen Dynasty through 1960s', *Journal of Social Sciences and Humanities* (Seoul), 26 (June 1967), 1–33.

CHOI JOO-WHAN, 'Estimates of North Korea's GNP', *The Unification Policy* (Seoul), 4:1 (1978).

CHOY BONG-YOUN, *Korea: A History* (Rutland, Vt., and Tokyo: Charles E. Tuttle, 1983).

——*A History of the Korean Reunification Movement: Its Issues and Prospects* (Peoria, Ill.: Bradley University, 1984).

CHUNG, JOSEPH S., 'The Six-Year Plan of North Korea: Targets, Problems and Prospects', *Journal of Korean Affairs* (Seoul), 1:1 (1971), 15–26.

CHUNG, JOSEPH S., *The North Korean Economy: Structure and Development* (Stanford, Calif.: Hoover Institution Press, 1974).

——'A Study of North Korean Economy', *Korea and International Politics*, 5:2 (Kyungnam University, 1989).

Chung-Ang Nyun-Gam (North Korea's Central Annual Report) (Pyongyang, 1949–1990).

CLARK, COLIN, *The Conditions of Economic Progress*, 3rd edn. (New York: Macmillan, 1957).

CLOUGH, RALPH N., *Embattled Korea: The Rivalry for International Support* (Boulder, Colo.: Westview Press, 1987).

CONROY, HILARY, *The Japanese Seizure of Korea, 1876–1910* (Philadelphia: University of Pennsylvania Press, 1960).

CUMMINGS, BRUCE, *The Origins of the Korean War: Liberation and the Emergence of Separate Regimes 1945–1947* (Princeton, NJ: Princeton University Press, 1981).

Current History Magazine (various issues).

DORNBUSH, RUDIGER, 'From Stabilization to Growth', paper presented at World Bank Annual Conference on Development Economics, Washington, DC, 26–7 April 1990.

DUNK, TIM, 'Prospects for Inter-Korea Cooperation', paper presented at the National Korean Studies Centre seminar 'Korea to the Year 2000', Seoul National University, 7 July 1992.

——'Two Koreas: Total Reunification?' paper presented at the seminar 'Korea to the Year 2000', Seoul National University, 7 July 1992.

Economic and Statistical Information on North Korea, Joint Publications Research Service 901-D: (15 January 1960), 122.

Economic Dictionary (Pyongyang: Social Science Publishing Co., 1970).

Economic Planning Board (EPB), *Social Indicators in Korea*, various issues.

Economist (The), *Book of Vital World Statistics* (New York: Random House, 1990).

Economist Intelligence Unit (EIU), *Country Report: China and North Korea*, various issues.

ETZIONI, AMITAI, *Eastern Europe: The Wealth of Lessons* (Washington, DC: George Washington University Press, 1991).

Europa World Yearbook, ii (London: Europa Publications, 1989).

FAO *Production Yearbook* (New York: UN, 1985).

FAO, *Monthly Bulletin of Statistics*, various issues (New York: United Nations, 1987).

FBIS, *Daily Report: East Asia*, 1 June 1990; 26 March 1990; 25 May 1990; 8 June 1990.

FELDSTEIN, M., 'Government Deficits and Aggregate Demand', *Journal of Monetary Economics*, 9 (1982), 1–20.

FRANK, CHARLES R., Jr, KWANG-SUK KIM, and LARRY E. WESTPHAL, *Foreign Trade Regimes and Economic Development: South Korea* (New York: Columbia University Press, 1975).

GONZALEZ, RODOLFO A., and STEPHEN L. MEHAY, 'Publicness, Scale, and Spillover Effects', *Public Finance Quarterly*, 18:3 (1990).

GRAJDANZEV, ANDREW J., *Modern Korea* (Seoul: Royal Asiatic Society, 1975).

HAGAN, EVERETT E., *The Economics of Development* (Homewood, Ill.: Richard D. Irwin, 1968; 2nd edn. 1975).

HALLIDAY, JON, 'The Economics of North and South Korea', in John Sullivan and Robert Foss (eds.), *Two Koreas—One Future?* (University Press of America, 1989).

HENDERSON, GREGORY, 'Korea', in G. Henderson *et al.* (eds.), *Divided Nations in a Divided World* (New York: David McKay, 1974).

HICKS, J. R., 'Income', in R. H. Parker and G. C. Harcourt (eds.), *Readings in the Concept and Measurement of Income* (Cambridge University Press, 1969).

HOLESOVSKY, VACLAV, *Economic Systems: Analysis and Comparison* (New York: McGraw-Hill, 1977).

HWANG EUI-GAK, 'The Macroeconomic Effects of Government Spending in South Korea, 1971–1988' (in Korean), *Korean Economic Studies* (Seoul) 4:1 (1990) (Korean Economic Research Institute, Seoul).

——'Trade Policy Issues between South Korea and the United States, with Some Emphasis on the Koreas' Position', paper presented at the Academic Symposium on the Impact of Recent Economic Developments on US–Korean Relations and the Pacific Basin, University of California, San Diego, 9–10 November 1990.

IYER, PICO, 'North Korea: In the Land of the Single Tune', *Time Magazine* (26 November 1990), 49–50.

JUNG HAE-KOO, 'The Bond Financing Effect of Budget Deficits and the Optimum Government Bond Financing', Ph.D. dissertation (Korea University, 1989).

KAKWANI, NANAK C., *Income Inequality and Poverty: Methods of Estimation and Policy Application*, a World Bank publication (Oxford University Press, 1980).

KIM, C. I. EUGENE and HAN-KYO KIM, *Korea and the Politics of Imperialism, 1876–1910* (Berkeley and Los Angeles: University of California Press, 1967).

KIM HA-KWANG, 'Pride in Having Superior Economic Management System which Gives Powerful Impetus to Consolidation and Development of Socialist System', *Rodong Sinmun* (20 December 1989).

KIM IL-SUNG, *Selective Work Book*, no. 4 (Pyongyang, 1960).

——*Theses on the Socialist Rural Questions in Our Country* (Pyongyang, 1964).

KIM KWANG-SUK and MICHAEL ROMER, *Growth and Structural Transformation* (Cambridge, Mass.: Harvard University Press, 1979).

KIM YOUNG-KYU, 'Methods of North Korea's GNP Estimates', *The Unification Policy* (Seoul), 6:3–4 (1980).

KIM YOUN-SOO (ed.), *The Economy of the Korean Democratic People's Republic* (Kiel: German–Korean Studies Group, 1979).

KNIGHT, PETER T., 'Economic Reform in Socialist Countries: The Experiences of China, Hungary, Romania and Yugoslavia', World Bank Staff Working Paper no. 579 (Washington, DC, 1983).

KOH BYUNG-CHUL, 'Unification Policies and North–South Relations', in R. A. Scalapino and J.-Y. Kim (eds.), *North Korea Today* (New York: Praeger, 1963).

KOMAKI, TERUO, 'Current Status and Prospects of the North Korean Economy', in M. Okonugi (ed.), *North Korea at the Crossroads* (Tokyo: Japan Institute of International Affairs, 1988).

KOMIYA, RUTARO, and KAZUO YASUI, *Japan's Macroeconomic Performance since the First Oil Crisis: Review and Appraisal*, Carnegie-Rochester Conference Series on Public Policy no. 20 (Amsterdam: North-Holland, 1984).

Korea Central News Agency (KCNA) (Pyongyang), 'Many Popular Welfare Measures in Force in North Korea' (21 March 1990).

Korea Herald (Seoul), 21 October 1990; 4 December 1990.

Korea Times (Chicago), 19 November 1985.

Korea Times (Seoul), 4 December 1990.

Korean Overseas Information Services, *A Handbook of Korea* (Seoul, 1988).

KORMENDI, R. C., 'Government Debt, Government Spending, and Private Sector Behavior', *American Economic Review*, 73 (1983), 994–1010.

KUARK, T., 'North Korea's Political Development During the Post-War Period', in Robert A. Scalapino and J. Kim (eds.), *North Korea Today* (New York: Praeger, 1963).

KUZNETS, PAUL W., *Economic Growth and Structure in the Republic of Korea* (New Haven, Conn.: Yale University Press, 1977).

——'Planning in Korea', paper presented at a conference on 'Indicative Planning', Brookings Institution, Washington, DC, April 1990.

Kyongje Konsul (Economy Construction) (Pyongyang), 2 (1990), 230–7.

LEE CHONG-SIK, 'Historical Setting', in William Evans-Smith (ed.), *North Korea* (Washington, DC: American University Press, 1981).

LEE HY-SANG, 'North Korea's Closed Economy: The Hidden Opening', *Asian Survey* (University of California at Berkeley) (December 1988), 1272.

LEE KI-PAEK, *Hankuk-sa Sinron* (Seoul: Ilchokak 1984).

LEE, PONG S., 'An Estimate of North Korea's National Income', *Asian Survey*, 12:6 (University of California Press, 1972), 518–21.

LEE, POONG, *Methods of GNP Estimates of North Korea* (Seoul: North Korea Research Institute, 1981).

LEIBFRITZ, WILLI, 'Economic Consequences of German Unification', *Business Economics*, 25:4 (1990), 7.

LELOUP, LANCE T., *Budgetary Politics: Dollars, Deficits, Decisions* (Ohio: King's Court Communications, 1977).

LEWIS, W. ARTHUR, *Development Planning: The Essentials of Economic Policy* (New York: Praeger, 1966).

MANSBRIDGE, JANE J. (ed.), *Beyond Self-Interest* (Chicago: University of Chicago Press, 1990).

MARER, PAUL, *Dollar GNPs of the USSR and Eastern Europe* (Baltimore: Johns Hopkins University Press, 1985).

MASON, EDWARD S., MAHN-JE KIM, DWIGHT H. PERKINS, KWANG-SUK KIM, and DAVID C. COLE, *The Economic and Social Modernization of the Republic of Korea* (Cambridge, Mass.: Harvard University Press, 1980).

MCCUNE, GEORGE M., *Korea Today* (Cambridge, Mass.: Harvard University Press, 1950).

MERRILL, JOHN, 'North Korea's Halting Efforts at Economic Reform', paper presented to the Fourth Conference on North Korea, Seoul, 7–11 August 1989.

MILLER, M., A. HOLM, and T. KOLLOHOR, 'Tumen River Area Development Mission Report', report of the UN Development Programme presented in Pyongyang, North Korea, 16–18 October 1991.

MORROW, ROBERT S., and KENNETH H. SHERPER, *Land Reform in South Korea* (Seoul: USAID/Korea, 1970).

MURREL, PETER, *The Nature of Socialist Economics: Lessons from Eastern European Foreign Trade* (Princeton, NJ: Princeton University Press, 1990).

National Unification Board (NUB), *A Study of North Korea's Agriculture* (Seoul: NUB, 1989).

——*Survey of North Korean Economy* (Seoul: NUB, 1988, 1989, 1990).

NORDHAUS, WILLIAM, 'The Longest Road: From Hegel to Haggle', paper presented at the Brookings Panel on Economic Activity, Washington, DC, 5–6 April 1990.

North Korea News (Seoul), 5 November 1990, 4–5.

North Korea's Statistics (1946–1985) (Seoul: National Unification Board, 1986).

OFER, GUR, 'Budget Deficit, Market Disequilibrium and Soviet Economic Reforms', *Soviet Economy*, 5 (April–June 1989), 107–61.

OKUN, ARTHUR, *Equality and Efficiency: The Big Trade-Off* (Washington, DC: Brookings Institution, 1975).

PARK, CHUNG-HEE, *To Build a Nation* (Washington: Acropolis Books, 1971).

PARK, HYUN-CHE, *Korean Economy and Agriculture* (in Korean) (Seoul: Gachi Publication Co., 1983).

PHRYMEA, PHOEBUS, *Econometrics: Stratified Foundations and Application* (New York: Harper & Row, 1970).

POND, ELIZABETH, *After the Wall: American Policy toward Germany*, a Twentieth Century Fund Paper (New York: Priority Press Publications, 1980).

ROBINSON, JOAN, 'Korean Miracle', *Monthly Review* (London) (January 1965).

Rodong Sinmun (Pyongyang). Each issue.

——'Let Us Effect Greater Upsurges in Producing People's Consumer Goods' (15 November 1990).

SALISBURY, HARRISON, *To Peking and Beyond: A Report on the New Asia* (New York: Quadrangle, 1973).

SCALAPINO, ROBERT A., and JUN-YOP KIM, *North Korea Today* (New York: Praeger, 1963).

SHERWOOD, ROBERT E., *Roosevelt and Hopkins: An Intimate History* (New York: Council on Foreign Relations, 1956).

STEINBURG, DAVID I., *The Republic of Korea: Economic Transformation and Social Change* (Boulder, Colo.: Westview Press, 1989).

SU TUNG-PAO, *Jun-Jokbyokboo* (pre-Red Wall thought), Sung Dynasty of China: AD 1036–1101.

SUH SANG-CHUL, 'Growth and Structural Changes in the Korean Economy since 1910', Ph.D. dissertation (Harvard University, 1966).

TEWKSBURY, DONALD G., *Source Materials on Korean Politics and Ideologies* (New York: Institute of Pacific Relations, 1950).

Tong-A Ilbo (Seoul), 14 October 1990.

US Department of State, *The Record on Korean Unification, 1943–1960* (Washington, DC: US Government Printing Office, 1960).

——'Foreign Relations of the United States: Diplomatic Papers, Conference at Cairo and Teheran, 1943' (Washington, DC: US Government Printing Office, 1961).

WAGNER, R. E., *Public Finance* (Boston: Little Brown, 1983).

WOLF, CHARLES, Jr, 'A Theory of Nonmarket Failure: Framework for Implementation Analysis', *Journal of Law and Economics*, 22 (April 1979), 107–39.

WOLF, THOMAS, 'Exchange Rates, Foreign Trade Accounting and Purchasing Power Parity for Centrally Planned Economies', World Bank Staff Working Papers no. 779 (1985).

World Development Report (Oxford: World Bank, 1985, 1986, 1987, 1988, 1989, 1990).

YEON HA-CHUNG, *North Korea's Economic Policy and Management* (Seoul: Korea Development Institute, 1986).

——'Bridging the Chasm', Korea Development Institute Working Paper no. 9012 (August 1990).

YU CHIN-O, 'Turning Period and Spiritual Direction', in *Direction of the New Generation* (in Korean) (Seoul: Hak-Won Sa, 1964).

Index